Evolving Complexity and Environmental Risk in the Prehistoric Southwest

EVOLVING COMPLEXITY AND ENVIRONMENTAL RISK IN THE PREHISTORIC SOUTHWEST

Proceedings of the Workshop
"Resource Stress, Economic Uncertainty, and Human Response in the Prehistoric Southwest,"
Held February 25-29, 1992 in Santa Fe, NM

Editors

Joseph A. Tainter
U.S. Department of Agriculture, Forest Service

Bonnie Bagley Tainter

Proceedings Volume XXIV

Santa Fe Institute
Studies in the Sciences of Complexity

Addison-Wesley Publishing Company
The Advanced Book Program

Reading, Massachusetts Menlo Park, California New York
Don Mills, Ontario Wokingham, England Amsterdam Bonn
Sydney Singapore Tokyo Madrid San Juan
Paris Seoul Milan Mexico City Taipei

Publisher: *David Goehring*
Executive Editor: *Jeff Robbins*
Production Manager: *Michael Cirone*
Production Assistant: *Amy Raufman Hannum*

Director of Publications, Santa Fe Institute: *Ronda K. Butler-Villa*
Publications Assistant, Santa Fe Institute: *Della L. Ulibarri*

This volume was typeset using TEXtures on a Macintosh II computer. Camera-ready output from a Hewlett Packard Laser Jet 4M Printer.

ISBN 0-201-87039-8 (Hardcover)
ISBN 0-201-87040-1 (Paperback)

1 2 3 4 5 6 7 8 9 10-MA-9998979695
First printing, December 1995

About the Santa Fe Institute

The *Santa Fe Institute* (SFI) is a multidisciplinary graduate research and teaching institution formed to nurture research on complex systems and their simpler elements. A private, independent institution, SFI was founded in 1984. Its primary concern is to focus the tools of traditional scientific disciplines and emerging new computer resources on the problems and opportunities that are involved in the multidisciplinary study of complex systems—those fundamental processes that shape almost every aspect of human life. Understanding complex systems is critical to realizing the full potential of science, and may be expected to yield enormous intellectual and practical benefits.

All titles from the *Santa Fe Institute Studies in the Sciences of Complexity* series will carry this imprint which is based on a Mimbres pottery design (circa A.D.950–1150), drawn by Betsy Jones. The design was selected because the radiating feathers are evocative of the outreach of the Santa Fe Institute Program to many disciplines and institutions.

Santa Fe Institute Editorial Board
June 1993

Santa Fe Institute
Studies in the Sciences of Complexity

Proceedings Volumes

Vol.	Editor	Title
I	D. Pines	Emerging Syntheses in Science, 1987
II	A. S. Perelson	Theoretical Immunology, Part One, 1988
III	A. S. Perelson	Theoretical Immunology, Part Two, 1988
IV	G. D. Doolen et al.	Lattice Gas Methods for Partial Differential Equations, 1989
V	P. W. Anderson, K. Arrow, D. Pines	The Economy as an Evolving Complex System, 1988
VI	C. G. Langton	Artificial Life: Proceedings of an Interdisciplinary Workshop on the Synthesis and Simulation of Living Systems, 1988
VII	G. I. Bell & T. G. Marr	Computers and DNA, 1989
VIII	W. H. Zurek	Complexity, Entropy, and the Physics of Information, 1990
IX	A. S. Perelson & S. A. Kauffman	Molecular Evolution on Rugged Landscapes: Proteins, RNA and the Immune System, 1990
X	C. G. Langton et al.	Artificial Life II, 1991
XI	J. A. Hawkins & M. Gell-Mann	The Evolution of Human Languages, 1992
XII	M. Casdagli & S. Eubank	Nonlinear Modeling and Forecasting, 1992
XIII	J. E. Mittenthal & A. B. Baskin	Principles of Organization in Organisms, 1992
XIV	D. Friedman & J. Rust	The Double Auction Market: Institutions, Theories, and Evidence, 1993
XV	A. S. Weigend & N. A. Gershenfeld	Time Series Prediction: Forecasting the Future and Understanding the Past
XVI	G. Gumerman & M. Gell-Mann	Understanding Complexity in the Prehistoric Southwest
XVII	C. G. Langton	Artificial Life III
XVIII	G. Kramer	Auditory Display
XIX	G. Cowan, D. Pines, and D. Meltzer	Complexity: Metaphors, Models, and Reality
XX	D. H. Wolpert	The Mathematics of Generalization
XXI	P. E. Cladis & P. Palffy-Muhoray	Spatio-Temporal Patterns in Nonequilibrium Complex Systems
XXII	H. Morowitz & & J. L. Singer	The Mind, The Brain, and Complex Adaptive Systems
XXIII	B. Julesz & I. Kovács	Maturational Windows and Adult Cortical Plasticity
XXIV	J. A. Tainter & B. B. Tainter	Evolving Complexity and Environmental Risk in the Prehistoric Southwest
XXV	J. B. Rundle, W. Klein & D. L. Turcotte	Reduction and Predicability of Natural Disasters
XXVI	R. K. Belew & M. Mitchell	Adaptive Individuals in Evolving Populations: Models and Algorithms

Lectures Volumes

Vol.	Editor	Title
I	D. L. Stein	Lectures in the Sciences of Complexity, 1989
II	E. Jen	1989 Lectures in Complex Systems, 1990
III	L. Nadel & D. L. Stein	1990 Lectures in Complex Systems, 1991
IV	L. Nadel & D. L. Stein	1991 Lectures in Complex Systems, 1992
V	L. Nadel & D. L. Stein	1992 Lectures in Complex Systems, 1993
VI	L. Nadel & D. L. Stein	1993 Lectures in Complex Systems, 1995

Lecture Notes Volumes

Vol.	Author	Title
I	J. Hertz, A. Krogh, & R. Palmer	Introduction to the Theory of Neural Computation, 1990
II	G. Weisbuch	Complex Systems Dynamics, 1990
III	W. D. Stein & F. J. Varela	Thinking About Biology, 1993

Reference Volumes

Vol.	Author	Title
I	A. Wuensche & M. Lesser	The Global Dynamics of Cellular Automata: Attraction Fields of One-Dimensional Cellular Automata, 1992

Contributors to This Volume

Eric A. Angstadt-Leto,
Arizona State University

Linda Cordell,
University Museum, University of Colorado

Jeffrey S. Dean,
Laboratory of Tree-Ring Research, University of Arizona

Michelle Hegmon,
New Mexico State University

Timothy A. Kohler,
Washington State University, Pullman and the Santa Fe Institute

Paul E. Minnis,
University of Oklahoma

Margaret C. Nelson,
State University of New York, Buffalo

Alison E. Rautman,
Michigan State University

Katherine A. Spielmann,
Arizona State University

Alan P. Sullivan III,
University of Cincinnati

Joseph A. Tainter,
U.S. Department of Agriculture, Forest Service

Carla R. Van West,
Crow Canyon Archaeological Center, Cortez, and Statistical Research, Inc.

For Elizabeth and George Tainter,
and in memory of Willis H. Bagley

Contents

Joseph A. Tainter
USDA Forest Service, Rocky Mountain Forest and Range Experiment Station, Albuquerque, NM 87106[1]

Introduction: Prehistoric Societies as Evolving Complex Systems

> One of the great challenges of contemporary science is to trace the mix of simplicity and complexity, regularity and randomness, order and disorder up the ladder from elementary particle physics and cosmology to the realm of complex adaptive systems.
>
> — Murray Gell-Mann (1994:119–120)

The late Southwestern archaeologist Emil Haury once asked a group of graduate students about their interests. When one aspiring archaeologist answered "complex societies," Haury queried, "Do you know of any simple societies?"[2] Human societies are by their nature among the most suitable subjects for the study of complexity.

[1]The preparation of this chapter was funded by the USDA Forest Service, Rocky Mountain Forest and Range Experiment Station. For comments on previous versions I am grateful to Linda Cordell, George Gumerman, David Kelley, Jane Kelley, Timothy Kohler, Richard Periman, Carol Raish, and Bonnie Bagley Tainter.

[2]I am grateful for this anecdote to my colleague Randall McGuire, who related it in a plenary address to the 23rd Annual Chacmool Conference at the University of Calgary, November, 1993. The theme of the conference was "Debating Complexity."

There being nothing simple in human social and cultural behavior, the topic offers two general problems that the study of complexity should account for. The first is the development of extrasomatic systems of problem solving, including technology, social relations, symboling, and language, among our hominid ancestors. These are among the constituents of what anthropologists call *culture*. The second is the tendency of human societies to change and vary in complexity, often very rapidly. These matters consume much of the attention of archaeologists, and should also concern scientists who study complexity across different types of systems.

This volume reflects the continuing interest of the Santa Fe Institute (SFI) in cultural complexity. It contains papers presented in a workshop titled *Resource Stress, Economic Uncertainty, and Human Response in the Prehistoric Southwest*, held 25–29 February 1992 at the Institute. The conference was planned by Linda Cordell, Marcus Feldman, Murray Gell-Mann, George Gumerman, and the author.[3]

This workshop was the fourth in a series. The series progressed from analysis of specific aspects of Southwestern prehistory, to broad theoretical formulations, and back again to specifics. The first conference was an Advanced Seminar held at the School of American Research (SAR), Santa Fe, in September 1983 under the title *Dynamics of Southwestern Prehistory* (Cordell and Gumerman 1989a). In this conference, Southwestern prehistory was approached from the perspective of subregions. Cordell and Gumerman (1989b) introduced the concept of *hinge points* in Southwestern prehistory. These were times of significant, rapid cultural change across much of the Southwest, distinguishable from the "background" pattern of relative stasis or localized change. The idea may apply to much of world prehistory and history, and is reminiscent of the biological concept of punctuated equilibrium (Gould and Eldridge 1977).

The first conference was followed by two more workshops conceived and organized by Gumerman and Gell-Mann, and sponsored jointly by SAR and SFI.[4] These workshops were intended to be a set, with the first establishing a conceptual foundation for the second. In September 1989, SAR held an Advanced Seminar titled *The Organization and Evolution of Prehistoric Southwestern Society*. The

[3]The workshop was funded by the USDA Forest Service, Rocky Mountain Forest and Range Experiment Station. Dr. George Peterson arranged the contractual matters for Rocky Mountain Station. The workshop participants included Linda Cordell, Pamela Bumstead, Jeffrey Dean, Marcus Feldman, Murray Gell-Mann, George Gumerman, Michelle Hegmon, Stuart Kauffman, Timothy Kohler, Robert Leonard, Paul Minnis, Margaret Nelson, Robert Preucel, Alison Rautman, Katherine Spielmann, Alan Sullivan III, Christine Szuter, Wolfgang Fikentscher, and Joseph Tainter. Andi Sutherland and Patrisia Brunello made our visit to SFI a most enjoyable experience. On behalf of all the authors, I am pleased to express appreciation to Ronda Butler-Villa and Della Ulibarri for their fine work in preparing the book for publication.

[4]Murray Gell-Mann's lifelong interest in archaeology, documented in his recent book (1994), has shaped the interest of the Santa Fe Institute and its affiliated scientists in Southwestern societies as complex adaptive systems. The two workshops on Southwestern prehistory held at SFI were made possible through his interest.

use of the singular form in the final word of the title suggests the concept behind both the seminar and the resulting book (Gumerman 1994a). While the *Dynamics of Southwestern Prehistory* conference mainly concerned approaches to local prehistory, this seminar emphasized evolutionary processes that occurred across much of the Southwest. These processes involved environmental and demographic stresses, hunter-gatherer land-use patterns, health and disease, aggregation, abandonment, and regional interaction. Gumerman (1994b) and Gumerman and Gell-Mann (1994a) introduced the volume with integrating syntheses.

The SFI workshop under the same title followed in October 1990. In SFI style it involved an assortment of scholars from various disciplines, working with archaeologists. These scientists had available the papers from the SAR Advanced Seminar, position papers drafted for the workshop[5] (Gumerman, Gell-Mann, and Cordell 1994:9–10), and introductory plenary addresses. Groups of archaeologists and other scientists worked on the focus topics for five days, producing a unique volume that merits a place in the history of archaeology (Gumerman and Gell-Mann 1994b). It is a blend of the perspectives of archaeology, ethnology, computational science, evolutionary biology, human physiology, and complexity theory, focused on understanding the evolution of prehistoric Southwestern societies as complex adaptive systems. It is the kind of merging of ideas and perspectives that SFI was established to bring about.

The workshop reported in this volume was a logical sequel to the previous efforts. The books resulting from the SAR Advanced Seminars and the SFI workshop brought together recent and thoughtful theories on the evolution of the prehistoric Southwest. In any field, though, the beauty and power of broad syntheses are realized fully when they help to clarify more specific topics: the devil is always in the details. The papers in this volume, taken together, provide a comprehensive view of the ways in which prehistoric Southwesterners made decisions and took steps to solve some of their everyday problems, and changed thereby the complexity of their economies, technologies, societies, and religious institutions. The cumulative total of such changes, many of which were reversible only under great hardship, comprises the evolution of Southwestern societies from small foraging bands to sedentary pueblo communities and regional networks. The strategies and decision criteria explored here are the mechanisms by which subsistence agricultural societies increase or decrease in complexity.

This introduction is intended for both archaeologists and other scientists interested in complexity. For the latter readers I will try not to plunge the discussion into archaeological technicalities. Instead I refer readers to the excellent overviews and syntheses to be found in Cordell (1984), Cordell and Gumerman (1989b), Gumerman (1994b), Gumerman and Gell-Mann (1994a), and Lekson, Cordell, and Gumerman (1994). In the next section I will discuss some topics of significance

[5]The position papers covered archaeological explanation, historical processes, environmental modeling, systems modeling, and environment, demography, and health.

in understanding cultural complexity.[6] Cultural complexity is different in some respects from complexity in other living systems, and is perhaps more enigmatic. The final section describes how the adaptations to risk and stress described in the various chapters provide an extensive glimpse into some of the ways that societies practicing subsistence economies change in complexity.

CULTURAL COMPLEXITY

COMPLEXITY AS TEXT

As every serious student of the topic knows, complexity can be quite difficult to define. While this is a disconcerting state of affairs for the term that identifies an entire field of learning, it is not unusual. Fundamental terms throughout science, such as *species*, *evolution*, *culture*, or *collapse*, have proved equally elusive. In the face of this problem it is tempting at times to adopt a Supreme Court type of concept: we may not be able to define complexity, but we know it when we see it.[7] Or at least we think we do. Recently I suggested a know-it-when-you-see-it illustration of cultural complexity that is worth presenting here (Tainter 1995b).

In the foothill country of southwestern Colorado, the U.S. Bureau of Land Management maintains a new institution of archaeological research and public interpretation: the Anasazi Heritage Center. It was built following several seasons of research among archaeological sites now inundated by McPhee Reservoir.[8] The Center rises behind a small pueblo ruin of the twelfth century A.D., the Dominguez Ruin. This ruin has been left uncovered and stabilized so that visitors can see an Anasazi ruin near the museum (Figure 1).

The juxtaposition of the two structures is illustrative. Here is a prehistoric pueblo, expediently built, consisting of a few rooms, and once home to a handful of people. The structure is small, and the architecture repetitive and predictable. Behind it rises a great edifice, many times the size of the little pueblo. It represents a small part of our abilities in engineering and materials science. It exists because our national government commissioned it to be built, and pays each year for a permanent staff, energy to heat and cool the building, and a fleet of vehicles. We

[6] The perspective offered here is that of an anthropologist concerned with the evolution of cultural complexity, and considering that evolution partly in an economic framework. Recognizing the advantages of a plurality of views, inherent in SFI's philosophy, I offer this perspective to augment the current discussion, not to displace any part of it.

[7] I am paraphrasing the pronouncement of a learned justice of the court in a 1970s case which concerned, in part, attempts to define pornography.

[8] The paper by Timothy Kohler and Carla Van West in this volume is an effort that grew out of this research.

FIGURE 1 The Simple and the Complex: Dominguez Ruin and the Anasazi Heritage Center. Dominguez Ruin is in the left foreground. Photograph by J. Fleetman, courtesy of Victoria Atkins and the U.S. Department of the Interior, Bureau of Land Management and Bureau of Reclamation.

undertake all this *merely to interpret* the small pueblo and others like it. Indeed, the energy we have spent to excavate these small pueblos, analyze and curate their remains, attend scientific conferences, publish interpretations and theories, and tell the public what we have learned may well exceed what the prehistoric Puebloans themselves consumed in their lives. Amassing and expending such large quantities of energy are hallmarks of a complex society (White 1949:363–393). Here, in the contrast between the Dominguez Ruin and the Anasazi Heritage Center, is surely a clear illustration of the difference between a society that was comparatively simple, in an anthropological sense, and one that is much more complex.

In an earlier study I advanced the following characterization of social complexity:

> Complexity is generally understood to refer to such things as the size of a society, the number and distinctiveness of its parts, the variety of specialized social roles that it incorporates, the number of distinct social personalities

> present, and the variety of mechanisms for organizing these into a coherent, functioning whole. Augmenting any of these dimensions increases the complexity of a society. Hunter-gatherer societies (by way of illustrating one contrast in complexity) contain no more than a few dozen distinct social personalities, while modern European censuses recognize 10,000 to 20,000 unique occupational roles, and industrial societies may contain overall more than 1,000,000 different kinds of social personalities [McGuire 1983:115] (Tainter 1988:23).

This characterization is along the lines of ones that other archaeologists have adopted (e.g., Plog 1974). It derives from the literature of systems theory (e.g., Miller 1965, 1978), to which many archaeologists were exposed in the 1960s and 1970s.

Gell-Mann has pointed out that while no objective definition of complexity has been found, in both scientific usage and popular discourse what is meant by the complexity of a system is essentially the length of the description of its regularities (Gell-Mann 1992, 1994; Gumerman and Gell-Mann 1994a). Complexity in this sense is not an attribute of a system but of our description of it. Complexity is not inherent in the object or system of perception. One hopes, of course, that there is (or can be) isomorphism between system complexity and the length of a description. Fortunately it does seem that other conceptions of complexity employed in archaeology are compatible with Gell-Mann's approach. In regard to the above quotation, for example, a society characterized by fewer parts, less differentiated parts, and fewer or simpler integrative mechanisms can certainly be described more succinctly than can a society with more of these. Dominguez Ruin can similarly be described more briefly than can the Anasazi Heritage Center. At least one anthropologist anticipated this approach over 40 years ago. Julian Steward pointed out the quantitative contrast between the 3,000 to 6,000 cultural elements documented by early ethnographers among native peoples of western North America, and the more than 500,000 artifact types landed by U.S. forces at Casablanca in World War II (1955:81).

Yet as Gell-Mann points out, even this conception (which we might call *complexity as text*) implies a plethora of difficulties. If complexity inheres only in descriptions, then it is likely to be context-dependent and subjective (Gell-Mann 1994:33). Differences in language, and in individual or cultural styles of communication, can lead to descriptions of different length being given to the same objective phenomenon. In the European intellectual tradition, for example, where the style of expository prose is often elliptical (e.g., Spengler 1962; Bourdieu 1984), the length of system descriptions may be longer than in other literary traditions. Presumably, then, comparisons of complexity should be based on the shortest message that *could* describe a system. That requires standardization of terminology and language, and equal levels of knowledge and skill among all communicants, things that are obviously infeasible. Archaeologists, for example, can hardly agree on a lexicon of stone-tool terminology, let alone standard descriptions of social features.

In a famous essay, Alfred Kroeber and Clyde Kluckhohn found a notorious level of diversity in definitions of the central concept of anthropology: culture (1952). Archaeologists cannot agree on the definition of what is an archaeological site (Tainter 1983).

Gell-Mann (1992, 1994) has analyzed in detail the idea of complexity as description, including such subtleties as crude complexity, algorithmic complexity, and lengths of schemata. Since this material is available to readers wishing to explore the topic further, I will turn to some implications for archaeology. Despite its subjectivity, the notion of complexity as text offers a perspective on archaeological research that practitioners may find useful. Some of the implications of that perspective merit brief remarks.

The apparent complexity of past cultural systems will have much to do with standards of fieldwork. It is possible that a Mesopotamian tell excavated poorly (by today's standards) in the nineteenth century could be described more succinctly than a pithouse village excavated well today. The urban society that produced the tell was undoubtedly more complex. As another example, our description of the complexity of a prehistoric land-use system will depend greatly on what we consider an archaeological deposit worthy of being recorded (Tainter 1979, 1983; Tainter and Lucas 1983). Archaeological data bases that exclude low-density remains or isolated artifacts will automatically lead to descriptions of past land uses that are shorter than would be appropriate.

The fact that a more complex society requires a longer description may be related in unexpected ways to how archaeologists set their research priorities, how they allocate time and other resources, and how they distribute their efforts among topics of study. In areas such as the Southwest or the Eastern Woodlands, the majority of archaeologists seem attracted to the periods of sedentary agricultural villages when populations were highest, cultures were most complex, the greatest numbers of sites were produced, and sites were most salient.[9] These periods accordingly have the most scientific literature. Fewer archaeologists specialize in either the preceding hunting and gathering periods (PaleoIndian and Archaic), or the period after European contact when populations declined. In both cases the cultural systems were, for the most part, simpler than in the intermediate periods, and sites tend to be fewer in number and less conspicuous. The level of archaeological effort that is expended on studying such simpler periods is suitable for cultural systems and their archaeological remains that will in the end require shorter descriptions. Phrased another way, Puebloan archaeology in the Southwest not only has more literature than PaleoIndian, Archaic, or Athabaskan archaeology, it *requires* more

[9] Salient archaeological sites are those that display the strongest patterning and that stand out most clearly from background noise (Tainter and Plog 1994). The most salient archaeological sites in the Southwestern United States are pueblos, and in the Midwest, burial, ceremonial, and residential mounds.

literature. Whatever the criteria by which archaeologists choose their research interests, those criteria, at least in these parts of North America, seem to result in a distribution of research effort that is serendipitously appropriate.[10]

Although human societies of the last 12,000 years or so have seemed inexorably to increase in complexity, this trend is interrupted occasionally by episodes of simplification. When major simplification occurs over a short period (roughly 50 to 100 years), it is considered a collapse (Tainter 1988:4). Some authors, citing aspects of cultural continuity across commonly recognized collapses (such as those of the Western Roman Empire or the southern Lowland Classic Maya), question the concept. Part of this confusion arises from the different meanings assigned to the term "collapse" (which, in different contexts, can mean such things as the demise of an empire, the consequence of a structural defect in a bridge, or what one does at the end of a difficult day). Another part of the confusion, though, arises from the fact that few social scientists attempt actually to measure complexity. The idea of complexity as text provides both a definition and a measure. By way of illustration, a colleague recently suggested to me that the complexity of the Roman Empire has been exaggerated relative to the complexity of the Germanic kingdoms that succeeded it in western Europe. If one considers the volumes of text describing these systems, such a notion seems spurious. The literature on the collapse of the Roman Empire began six centuries before the event itself (Polybius 1979), and has scarcely known an idle period since (Tainter 1988, 1994b). As for the literature on Merovingian Gaul or Visigothic Spain, the Dark Ages are called that with good reason. While the relative sizes of the literatures on these systems does not prove that the Roman Empire was more complex,[11] it does suggest that a system description of Rome requires more intensive scholarly effort than do descriptions of early post-Roman societies. It also suggests approaches to resolving the dispute.

A similar approach can be applied in the Southwest. Were there prehistoric Southwestern societies that were more complex than the historic Pueblos? How complex was the Chacoan system at A.D. 1050 vs. 1200 (i.e., pre- and post-collapse)? How does contemporary Piman society differ in complexity from that of the Hohokam? While there are no definitive answers to such questions, insight into the nature of these issues is gained merely by asking how long a complete description of each system would need to be. Even an intuitive response would improve the quality of some debates in Southwestern archaeology.

[10]This subjective impression may seem to be contradicted by the relative paucity of archaeologists practicing Euroamerican archaeology in North America. The resolution of this contradiction may lie in the fact that Euroamerican culture is already well known through historical records and our everyday experiences.

[11]It is possible that the relative sizes of these literatures reflect scholarly aversion to early Medieval Europe, but that seems unlikely.

DISTINCTIVENESS OF CULTURAL COMPLEXITY

The post-World War II literature on systems theory and information theory (e.g., Weaver 1949; Shannon 1949; Bertalanffy 1968; Gatlin 1972) in some ways anticipated today's interest in complex adaptive systems. One of the points established in this literature is that living systems are characterized by structural and processual regularities (e.g., Miller 1965, 1978). These regularities make it possible to generalize about complex adaptive systems. Yet inevitably, there will always be specialists who feel that their type of complex adaptive system is unique enough to merit special consideration. Anthropologists are certainly no exception to this. Notwithstanding this tendency to scientific niche separation, there are indeed aspects of cultural complexity that merit special discussion. The first of these, the *cost of complexity*, may be applicable to all living systems, though not in quite the same ways. The second, the *meaning of complexity*, is, as far as we know, exclusively human.

In the world of complex adaptive systems there is, to use a colloquial expression, no free lunch. Complexity always has an energy cost. As the complexity of an adaptive system increases, so also does the quantity[12] of energy needed to create, maintain, and replace the system's components, to support their interactions, and to regulate their behavior. Leslie White's ideas on the relationship of energy capture to the evolution of culture (1949:363–393) made it clear that energy and cultural complexity are opposite sides of a coin. He once estimated that a cultural system activated primarily by human energy can generate only about 1/20 horsepower per capita per year (White 1949:369, 1959:41–42). That is, moreover, all the energy that such a system requires. In societies today, 1/20 horsepower per capita suffices only for a fleeting moment of industrial life. Complexity costs.

A complex adaptive system, as characterized by Gell-Mann, "...acquires information about its environment and its own interaction with that environment, condensing those regularities into a kind of 'schema' or model, and acting in the real world on the basis of that schema" (1994:17). Human cognition and culture clearly qualify as complex adaptive systems (although not all anthropologists choose to treat them as such). In the evolution of cultures as adaptive systems, complexity has been a primary problem-solving tool. There is much literature, including the papers in this volume, to show that the risks and stresses that human societies have faced have often been resolved by becoming more complex. Whether this has been in the realm of technology, economics, settlement, sociopolitical organization, or information processing, as human populations have found existing arrangements at any time unsuitable, the solution has been to increase complexity in one or more of these dimensions (Tainter 1988, 1995a). The development of increasingly complex military technology and organization provides a particularly clear example of this

[12] Requirements for high quality energy may increase as well. For the concept of energy quality (the ability of different kinds of energy to support useful work), see Hall, Cleveland, and Kaufmann (1992:55–56).

(Tainter 1992), but there are others in the areas of subsistence (Boserup 1965; Clark and Haswell 1966; Asch, Ford, and Asch 1972; Cohen 1977), sociopolitical organization (Tainter 1988, 1994b), and information processing (Machlup 1962; Rescher 1978, 1980; Rostow 1980; Tainter 1988, 1995a). It is reasonable to suggest that the success of humans as a species is attributable not only to large and richly networked brains, upright posture, and opposable thumbs, but also to the fact that these attributes allow cultural systems rapidly to become more complex.

The development of cultural complexity is an economic process: complexity levies costs and yields benefits. No doubt this is true of all complex adaptive systems. Every structure and process in an organism, for example, has a metabolic cost. Any increase in the complexity of a language imposes steeper learning requirements and a greater chance of miscommunication.

If the development of complexity is an economic process, then the appearance of more complex cultural behavior must always imply a benefit/cost calculation. Such calculations have rarely been explicit in human history (though people seem intuitively to understand the concept), and in hierarchical societies those who benefit from complexity are often not those who must bear its costs. Nevertheless, the benefit/cost ratio to investment in complexity has powerfully influenced cultural evolution, and the course of human history (Tainter 1988, 1994a, 1994b).

The fact that complexity is a benefit/cost equation influences cultural evolution in at least two major ways. The first is that the cost of becoming more complex must always have tended to inhibit the development of cultural complexity. If people must work harder to support complex institutions (Tainter 1994a), why do so unless there is a clear need or benefit? This simple point clarifies major riddles in our history. For example, it helps us to understand why, although our evolution as a species extends over several million years, the most complex societies—states—have existed for only about five millennia. There is no latent or inherent tendency to cultural complexity, as many authors (and much of the public) have mistakenly assumed. Complexity is a problem-solving response.

The economic nature of cultural complexity influences human history in a second way: investment in increasing complexity can reach the point of diminishing returns. Developing costly institutions is suitable as long as there are stable or increasing returns to the investment. Ultimately, though, as inexpensive technological or organizational solutions are exhausted, increasing complexity reaches the point of declining marginal returns. Beyond this point growing more complex yields progressively lower benefits per unit of investment (that is, the marginal utility of further complexity declines). Complex societies that have reached this point have usually had three options: impoverish the support population, acquire new energy subsidies to pay for greater complexity (often accomplished in ancient societies by expanding territorially), or collapse. These options are not mutually exclusive: often the first and second lead ultimately to the third (Tainter 1988).

The problem of diminishing returns to complexity is well illustrated by the development and collapse of the Western Roman Empire. When confronted with a military crisis, which happened increasingly over time, Roman Emperors often found the money to respond by debasing the silver currency and seeking ways to raise new funds. When military crises became constant in the third century A.D., the Emperors doubled the size of the army and increased both the size and complexity of the government. To pay for this, masses of worthless coins were produced, supplies were commandeered from peasants, and the level of taxation was made even more oppressive. Lands and population were surveyed across the empire and assessed for taxes. Communities were held corporately liable for any unpaid amounts. As overtaxed peasants went hungry or sold their children into slavery, massive fortifications were built across the empire, the size of the bureaucracy doubled, provincial administration was made more complex, large subsidies in gold were paid to German tribes, and new imperial cities and courts were established. As taxes rose, marginal lands were abandoned and population declined: peasants could no longer support large families. To avoid now-oppressive civic obligations, the wealthy fled from cities to establish self-sufficient rural estates. Ultimately, to escape taxation, peasants voluntarily entered into feudal relationships with these land holders. A few wealthy families came to own much of the land in the western empire, and were able to defy the imperial government. The empire came to sustain itself by consuming its capital resources: producing lands and peasant population. Collapse was inevitable (Jones 1964, 1974; Tainter 1988, 1994b). The Roman Empire provides perhaps history's best-documented example of how increasing complexity to resolve problems leads to higher costs, diminishing returns, alienation of a support population, economic weakness, and collapse (Figure 2).

There is another aspect of cultural complexity that is uniquely human: *people give meaning to complexity.* We assign value to it. People care about how complex their lives are, and whether their government is worth the cost. No other complex adaptive system has this characteristic. Neither Darwin's finches, nor chimpanzees, nor, so far as we know, any other living system constructs symbols regarding the complexity of its behavior. We are the only species that can increase the complexity of its behavior, and then wonder if we were right to do so.

As a simple illustration of this point, consider the meaning of complexity in a Chinese banquet. The quantity, diversity, and complexity of the dishes, and how they are presented, are used not only to provide calories to the guests, but also to convey a variety of cultural meanings. These include the status of the host, the importance of the guests, and the significance of the occasion. Of course the use of complexity in this way is meaningful largely because the cost of the banquet varies with its complexity.

The power of our ability to give meaning to complexity should never be underestimated. The difference between an approach that incorporates this fact and one that ignores it is at least as great as the difference between Lamarckian and Mendelian inheritance. Many aspects of our behavior can be characterized as *complexity-averse.* In science, the Principle of Occam's Razor has enduring appeal because it states clearly that simplicity in explanation is preferable to complexity. The so-called "complexity of modern life" is a regular complaint in public discourse.

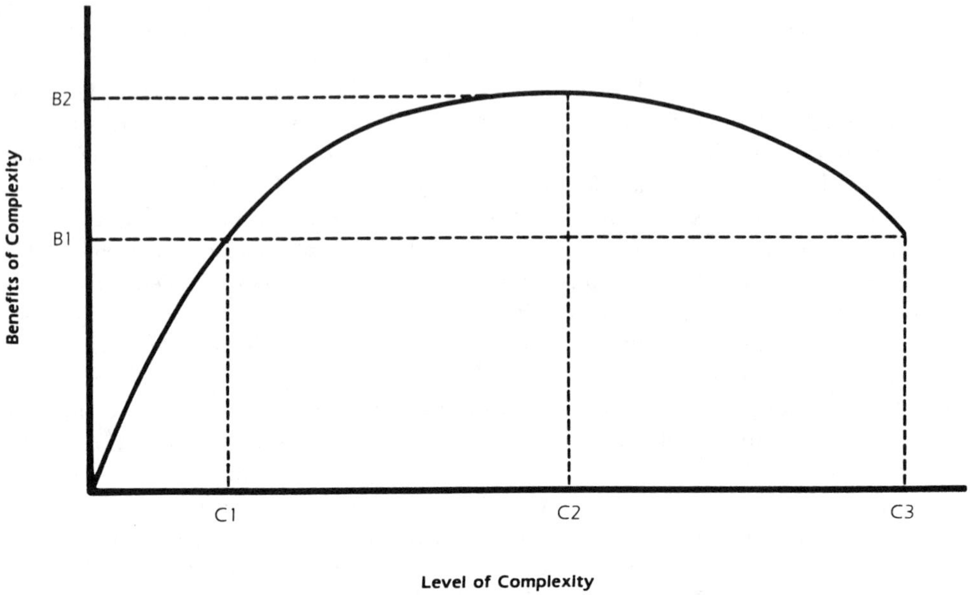

FIGURE 2 The Marginal Return to Investment in Complexity (after Tainter 1988:119). The area on the curve beyond B1,C1 marks a region of diminishing returns to complexity, and associated economic and political problems. At B1,C3 the benefits of social and political organization have declined to those available at lower levels of complexity and expenditure. For a society in such a condition collapse is imminent.

Much of the current popular discontent with government stems from the fact that government adds complexity to people's lives, through behavioral regulation and increases in the number and diversity of activities in which people must engage. So strong is the aversion to hierarchically imposed complexity that politicians in our day successfully base their careers on exploiting the discontent it creates, and journalists win prizes for exposing it.

Recently I had an opportunity to observe government-run environmental education projects in traditional villages in southern Mali. These efforts, and the reaction to them, exemplified the topics of this chapter: increasing complexity, increasing costliness of complexity, giving meaning to complexity, and aversion to complexity. The educational endeavors concerned such things as firewood use, soil productivity, and gathering honey. The education is done in the context of traditional types of village gatherings (although called by government officials) or by professional acting troupes (Figure 3). The basic message is that villagers should do things to enhance conservation. These include planting trees, and using improved types of beehives and stoves. Yet to do these things increases the complexity and costliness

FIGURE 3 Exhortations to Complexity: Environmental Education in Southern Mali. Photograph by the author.

of the villagers' lives. A villager who plants a tree must, when it is mature, obtain a government permit to cut it down. This requires money, which is always hard to come by, and a trip to the nearest forestry official. To obtain an improved beehive the villager must travel to the capital, Bamako, and have still more money. Not surprisingly, while villagers enjoy the gatherings and listen politely to the conservation message, they do not seem to rush to adopt the new technologies.

Development workers, and indeed most outsiders, would ascribe this to conservatism. Malians, like all people, are indeed emotionally attached to traditional ways of doing things, and change is stressful. Undoubtedly there is some of this in any resistance to change. There is another aspect to the matter, though: the traditional ways are simpler and less expensive. Even in the absence of emotional attachment, this would suffice to explain why new technologies are not embraced.

Malians give meaning to complexity in daily social relations. At rural markets, for example, government officials sit next to cartoonlike billboards that illustrate how the government recommends people manage their lives. (Since Mali has a literacy rate of about ten percent, the messages are conveyed through drawings, which the government official elaborates upon to anyone who approaches.) One billboard exhorts people to conserve their flocks of chickens: sell eggs and keep your laying hens. Villagers respond with statements like "Perhaps [the man depicted in the drawings] had to sell his chickens to buy medicine for a sick relative." In this answer, the complexity of social relations is employed to deny the government's claim to superior knowledge and to resist the government's exhortations. In Bamako itself, although residents always have rural kin who can supply much of their needs, they often deploy the complexity of a "Western" style of living to signify their "modernity." This extends even to how many wives and children a man wishes to have.[13] Malians can either adopt or deny complexity to indicate their views of Mali's desired future (Western or traditional) and to claim a place in this social order.

Further complicating the matter is the fact that most of this complexity—government instructors, education, new technologies, urban life—is supported by funds from extraterritorial donors. As Mali adopts increasing complexity it must increase its involvement with external financiers, who will seek to channel the nation's policies. In the case of a state like Mali, part of the price of financing complexity is still more complexity.

Cultural complexity thus shows some interesting paradoxes. It is both self-reinforcing and self-inhibiting. Complexity reinforces itself through several mechanisms, such as the following.

A. As the complexity of one part of a cultural system increases (e.g., technical specialization), other parts may also need to become more complex (e.g., economic and social integration) (e.g., Olson 1982; Tainter 1988).

B. Economic development may be needed to pay the cost of higher complexity.

C. Since complexity may confer military advantages (Tainter 1992), the neighbors of a more complex society may need to adapt by increasing their own complexity to a similar level.

In the contrary direction, complexity is inhibited by its cost and by complexity aversion. The latter may largely be a function of cost.

[13] Unfortunately I was not able to talk to Malian women about these matters. Not as many of them speak French as do men, and attempts to approach them would have been misunderstood. It is interesting, however, that during my first visit (October-November 1992) our translator published the first Western-style "women's magazine" to appear in Mali. Both the publishing and the reading of such a magazine signify "modernity," yet acquiring literacy and the money to buy such a magazine adds complexity to people's lives.

The negotiation of these conflicting forces creates tension in evolving complex societies.[14] This tension can be seen in the fact that while we value complex societies (calling them "civilizations" in popular discourse), people prefer not to pay the cost of complexity, and seek to minimize it in their own lives. Cultural complexity is qualitatively different from complexity in other kinds of living systems. Cultural complexity generates consciousness of itself, and this consciousness in turn modifies complexity. Compared to other kinds of systems, we may find that cultural complexity is (for a lack of a better term) more complex.

EVOLVING COMPLEXITY IN PREHISTORIC SOUTHWESTERN SOCIETIES

Cultural evolutioncultural complexity, change in (i.e., changes in cultural complexity) responds to a number of stimuli. The papers in this volume focus on one of these: the need to gain adequate energy from the natural environment. This is the topic the workshop participants were charged to address, and they were selected for having done significant research on the matter. Other scholars emphasize different stimuli for evolving complexity. The self-reinforcing and self-inhibiting nature of cultural complexity have just been discussed. In hierarchical societies, conflict between rulers and ruled has much to do with whether complexity changes, and in what direction. Complexity may increase from either trade or competition among equivalent societies, which Renfrew has labeled "peer polities" (1982, 1986; see also Price 1977; Tainter 1988, 1992). These are all valid perspectives, and should be components of a general model. In the arid Southwest, though, archaeologists have long found strategies of survival a compelling topic.

In her concluding assessment, Linda Cordell points out that definitions of stress or risk are absent from many papers, and operationally defined in others. Margaret Nelson is perhaps the most explicit in addressing issues of definition. It is appropriate to consider the matter here. In the context of adaptation to the natural environment, I consider *stress* to be the consequences that arise when the energy needs of a human population are not met. The term "energy needs" in this case includes the nutrients people must consume, energy for other physical needs (such as cooking and heating), and energy required to fund extrasomatic aspects of the cultural system (e.g., trade, ceremonies, prestige goods). *Risk* is the likelihood that energy needs will not be met,[15] and can be considered a set of probability distributions covering stresses of different durations and degrees of intensity. Short-duration

[14]It did so in prehistory as well. Alden Hays (1981) has suggested that contacts with Mesoamericans could have led to factionalism in Puebloan societies.

[15]Bruce Winterhalder's conception of risk is along similar lines (1990:67), and is adopted by Nelson (this volume).

or mild stress is *a priori* more likely than long-duration or severe stress. Conceived this way, there obviously is no simple point at which risk begins or stress sets in. Both are continuous distributions, and it is meaningful to talk only of degrees of stress and risk. Moreover, a risk that generates stress at one time, but to which the cultural system responds, may thereafter cease to be risky. An example would be the risk of starvation in a late winter/early spring lean season which is ameliorated by the development of storage technologies or exchange ties.

The papers in this volume cover a variety of topics. The study by Jeffrey Dean is fundamental to all of the others. Continuing his work on the effects of low-frequency (≥ 25 years) and high-frequency (< 25 years) environmental fluctuations, he traces the distribution of unimodal and bimodal annual precipitation patterns across the Southwest. The patterns remain the same for about 1500 years, a period long enough for cultural memory no longer to retain techniques for coping with departures from the pattern (see Gunn 1994). The distribution breaks down between A.D. 1250 and 1450, during which time there is no simple geographical pattern. This event came at the end of a period of population expansion (Gumerman and Gell-Mann 1994a:20–22), and coincides in time with one of the major, unexplained facts of Southwestern prehistory: the apparent abandonment of the Four Corners region and large parts of the uplands of Arizona and New Mexico.

Paul Minnis's paper should also be read as background to those that follow. Minnis's topic concerns responses to food acquisition problems, and the sequence in which different responses might be tried. He employs a scheme developed by Halstead and O'Shea (1989), who classify responses into mobility, diversification, physical storage, and exchange. Minnis suggests that responses that are less costly and more reversible will be employed before responses without these qualities. Importantly, he proposes that social relations provide an effective vehicle for responding to the most serious food provisioning problems, but their cost in reciprocal obligations is high.

Katherine Spielmann and Eric Angstadt-Leto consider the consequences of specific nutrient deficiencies. Strategies for countering shortfalls in meat include trade, turkey husbandry, and the use of plants with some of the nutrients that meat provides. Trade in meat is not well documented in most areas. Turkeys did become an important source of protein in some times and places. Except for beans, there is no evidence for great emphasis on plants with nutrients complementary to those found in maize. A lack of adequate animal protein may have been a recurrent source of stress on Puebloans.

Technological change is a primary response to a need for subsistence intensification. Margaret Nelson provides perhaps the archaeological literature's best description of how technological organization responds to subsistence stress. Possible responses include subsistence specialization, diversification, and pooling of risk. Designing technology to respond to risk is an appropriate strategy, but has costs. For example, designing a projectile point to stay in an animal may cost the point and shaft, or foreshaft. Tool makers have to balance the advantages and disadvantages of technological designs. An optimal tool form combines attributes that suit economic and social strategies.

Nelson's paper may be particularly useful to SFI scientists who wish to understand the archaeological record better. To an untrained eye, the stone tool debris on an archaeological site appears bewildering. This paper provides a framework to make sense of that variety.

Alan Sullivan's paper is a bold attempt to reshape our thinking about prehistoric Southwestern subsistence organization. Sullivan challenges the historical assumption that the Southwestern environment is innately hostile to humans who must make a living from it. He argues to the contrary, that prehistoric Southwesterners were fully capable of manipulating environmental productivity so that they would have experienced few gaps in food supply, or even none. Sullivan's ideas on food for the Anasazi provide, one might say, much food for thought. Questions about population size and trends, territoriality, and environmental degradation in relation to environmental manipulation will be significant topics for future research. Sullivan makes the important suggestion that we have misunderstood Puebloan subsistence because we have focused our research on locations where food was consumed (pueblos) rather than locations where it was produced. The latter may be small, ephemeral artifact scatters. Many archaeologists overlook the importance of these small sites (Tainter 1979, 1983).

Timothy Kohler and Carla Van West discuss food sharing as the foundation for village life. From a donor's perspective, sharing is attractive when production is beyond the point at which the marginal utility of production starts to diminish. Kohler and Van West project that pooling will be favorable during periods of high mean production coupled with high annual fluctuations in productivity, or high spatial variability. Periods of low mean production should discourage production beyond household needs, and encourage defection from pooling networks. In general the climatic and archaeological records of their study area in southwestern Colorado support the prediction. An unanticipated finding of this study is that high population levels also favor sharing.

It is worth considering this economic model in the context of the points made previously, that people assign meaning to complexity. This is especially so in social contexts. Sharing has meanings that go beyond calories, including kinship, marriage arrangements, exchange of information, and reinforcement of social ties. Hegmon's paper in this volume considers some of these factors.

Alison Rautman shifts the consideration of sharing from the level of community to region. Social interaction within a network provides for sharing of information about regional conditions, and a context within which to exchange resources. In central New Mexico, she finds, ceramic assemblages suggestive of interaction are strongest between areas that experience complementary rainfall distributions, and that in prehistory likely experienced different resource productivity patterns. She points out that exchanges in such networks cannot be balanced: a partner could simply withdraw. Some indebtedness is needed to encourage further participation.

Too much asymmetry in exchanges, however, is the stuff of which inequality is made. Thus the existence of an exchange system may generate further complexity in the social sphere.

Michelle Hegmon takes note of the fact that from ca. A.D. 400–1300, Anasazi reliance on maize did not greatly change, but many other cultural changes took place. These changes cannot be understood solely in terms of subsistence, but are intelligible in terms of social relations as part of modes of production. Maize is important in Puebloan life not only for food but also because people give it meaning: it has roles in social and religious life. Other foods can substitute for maize in subsistence, but not in social or religious contexts. In a simulation of variability and risk in Hopi agriculture, Hegmon finds that without sharing, fewer than 50 percent of households are likely to survive 20 years. With pooling the survival rate averages 72 percent, but one household that does poorly can pull down the entire group. Restricted sharing (sharing within groups of five households) increases the survival rate to 92 percent. This is similar to the strategy the Hopi actually use. Hegmon proposes that in the transition from pithouse-to-puebloan architecture, the strategy of storing food changed from one that was somewhat public to one that was mostly private.[16] The latter allows for restricted sharing, and is demonstrably beneficial to household survival. But it is associated with social inequality. Hegmon's results suggest that restricted sharing and social inequality are beneficial, and might have emerged consensually. This is significant: many social theorists assume that inequality is detrimental, and must have emerged through conflict. The debate is as old as inequality itself (Tainter 1988:22–38).

Linda Cordell concludes with an assessment that places the papers in the context of Southwestern archaeology and the interests of the Santa Fe Institute. She suggests that the fourteenth-century abandonment of much of the Southwest ranks as one of history's greatest failures, and as such merits broad scholarly attention.

The papers in this volume, then, provide a glimpse of how subsistence agricultural societies function as complex adaptive systems. In each dimension of adaptation—whether technology, subsistence, environmental manipulation, or social relations—complexity was employed in the prehistoric Southwest as a primary strategy to counter risk and ameliorate stress. Undeniably these strategies were successful, yet understanding that success is not the end of our research. Once cultural complexity exists, its cost and the meanings people give to it make it at least partly a thing *sui generis*. Cultural complexity modifies itself, which renders it tantalizing and elusive as a topic of research. The challenges to understanding cultural complexity mark it as a subject worthy of the Santa Fe Institute.

These papers concern prehistory, but in some ways things have not changed. Puebloans today use cultural complexity as part of their strategy to maintain cultural identity in the face of great pressures to assimilate. The connections among Puebloan ritual, social relations, community, agriculture, and cosmology form a coherent complex system that enables the Puebloans to resist unwanted change. Now

[16]This point has previously been made by Linda Cordell (1979:100–101).

as in the past, native Southwesterners employ complexity to survive physically, and to survive culturally.

REFERENCES

Asch, Nancy B., Richard I. Ford, and David L. Asch

1972 *Paleoethnobotany of the Koster Site: The Archaic Horizons.* Reports of Investigations No. 24. Illinois State Museum, Springfield. Illinois Valley Archeological Program, Research Papers No. 6. Center for American Archaeology, Kampsville.

Bertalanffy, Ludwig von

1968 *General System Theory.* George Braziller, New York.

Boserup, Ester

1965 *The Conditions of Agricultural Growth: The Economics of Agrarian Change Under Population Pressure.* Aldine, Chicago.

Bourdieu, Pierre

1984 *Distinction: A Social Critique of the Judgement of Taste* (translated by Richard Nice). Harvard University Press, Cambridge.

Clark, Colin, and Margaret Haswell

1966 *The Economics of Subsistence Agriculture.* MacMillan, London.

Cohen, Mark N.

1977 *The Food Crisis in Prehistory: Overpopulation and the Origins of Agriculture.* Yale University Press, New Haven.

Cordell, Linda S.

1979 *A Cultural Resources Overview of the Middle Rio Grande Valley, New Mexico.* USDA Forest Service, Southwestern Regional Office and USDI Bureau of Land Management, New Mexico State Office, Albuquerque and Santa Fe.

1984 *Prehistory of the Southwest.* Academic Press, New York.

Cordell, Linda S., and George J. Gumerman (editors)

1989a *Dynamics of Southwest Prehistory.* Smithsonian Institution Press, Washington, D.C.

Cordell, Linda S., and George J. Gumerman

1989b Cultural Interaction in the Prehistoric Southwest. In *Dynamics of Southwest Prehistory*, edited by Linda S. Cordell and George J. Gumerman, pp. 1–17. Smithsonian Institution Press, Washington, D.C.

Gatlin, Lila L.

1972 *Information Theory and the Living System.* Columbia University Press, New York.

Gell-Mann, Murray
1992 Complexity and Complex Adaptive Systems. In *The Evolution of Human Languages*, edited by John A. Hawkins and Murray Gell-Mann, pp. 3–18. Santa Fe Institute Studies in the Sciences of Complexity, Proceedings Volume XI. Addison-Wesley, Reading.
1994 *The Quark and the Jaguar: Adventures in the Simple and the Complex.* W. H. Freeman, New York.
Gould, Stephen J., and Niles Eldredge
1977 Punctuated Equilibria: The Tempo and Mode of Evolution Reconsidered. *Paleobiology* 3:115–151.
Gumerman, George J. (editor)
1994a *Themes in Southwest Prehistory.* School of American Research Press, Santa Fe.
Gumerman, George J.
1994b Patterns and Perturbations in Southwestern Prehistory. In *Themes in Southwest Prehistory*, edited by George J. Gumerman, pp. 3–10. School of American Research Press, Santa Fe.
Gumerman, George J., and Murray Gell-Mann
1994a Cultural Evolution in the Prehistoric Southwest. In *Themes in Southwest Prehistory*, edited by George J. Gumerman, pp. 11–31. School of American Research Press, Santa Fe.
Gumerman, George J., and Murray Gell-Mann (editors)
1994b *Understanding Complexity in the Prehistoric Southwest.* Santa Fe Institute Studies in the Sciences of Complexity, Proceedings Volume XVI. Addison-Wesley, Reading.
Gumerman, George J., Murray Gell-Mann, and Linda Cordell
1994 Introduction. In *Understanding Complexity in the Prehistoric Southwest*, edited by George J. Gumerman and Murray Gell-Mann, pp. 3–14. Santa Fe Institute Studies in the Sciences of Complexity, Proceedings Volume XVI. Addison-Wesley, Reading.
Gunn, Joel D.
1994 Global Climate and Regional Biocultural Diversity. In *Historical Ecology: Cultural Knowledge and Changing Landscapes*, edited by Carole L. Crumley, pp. 67–97. School of American Research Press, Santa Fe.
Hall, Charles A. S., Cutler J. Cleveland, and Robert Kaufmann
1992 *Energy and Resource Quality: The Ecology of the Economic Process.* University Press of Colorado, Niwot.
Halstead, Paul, and John O'Shea
1989 Introduction: Cultural Responses to Risk and Uncertainty. In *Bad Year Economics: Cultural Responses to Risk and Uncertainty*, pp. 1–7. Cambridge University Press, Cambridge.
Hayes, Alden C.
1981 A Survey of Chaco Canyon Archaeology. In *Archaeological Surveys of Chaco Canyon*, by Alden C. Hayes, David M. Brugge, and W. James

Judge, pp. 1–68. Publications in Archeology No. 18A. USDI National Park Service, Washington, D.C.

Jones, A. H. M.

1964 *The Later Roman Empire, 284–602: A Social, Economic and Administrative Survey.* University of Oklahoma Press, Norman.

1974 *The Roman Economy: Studies in Ancient Economic and Administrative History.* Basil Blackwell, Oxford.

Kroeber, Alfred, and Clyde Kluckhohn

1952 *Culture: A Critical Review of Concepts and Definitions.* Papers of the Peabody Museum of American Archaeology and Ethnology, Vol. 47. Harvard University, Cambridge.

Lekson, Stephen, Linda Cordell, and George J. Gumerman

1994 Approaches to Understanding Southwestern Prehistory. In *Understanding Complexity in the Prehistoric Southwest*, edited by George J. Gumerman and Murray Gell-Mann, pp. 15–24. Santa Fe Institute Studies in the Sciences of Complexity, Proceedings Volume XVI. Addison-Wesley, Reading.

Machlup, Fritz

1962 *The Production and Distribution of Knowledge in the United States.* Princeton University Press, Princeton.

McGuire, Randall H.

1983 Breaking Down Cultural Complexity: Inequality and Heterogeneity. In *Advances in Archaeological Method and Theory*, Volume 6, edited by Michael B. Schiffer, pp. 91–142. Academic Press, New York.

Miller, James G.

1965 Living Systems: Basic Concepts. *Behavioral Science* 10:193–237.

1978 *Living Systems.* McGraw-Hill, New York.

Olson, Mancur

1982 *The Rise and Decline of Nations.* Yale University Press, New Haven.

Plog, Fred

1974 *The Study of Prehistoric Change.* Seminar Press, New York.

Polybius

1979 *The Rise of the Roman Empire* (translation by Ian Scott-Kilvert of *The Histories*). Penguin, Harmondsworth.

Price, Barbara

1977 Shifts of Production and Organization: A Cluster Interaction Model. *Current Anthropology* 18:209–34.

Renfrew, Colin

1982 Polity and Power: Interaction, Intensification and Exploitation. In *An Island Polity: The Archaeology of Exploitation on Melos*, edited by Colin Renfrew and Malcolm Wagstaff, pp. 264–90. Cambridge University Press, Cambridge.

1986 Introduction: Peer Polity Interaction and Socio-Political Change. In *Peer Polity Interaction and Socio-Political Change*, edited by C. Renfrew and J. F. Cherry, pp. 1–18. Cambridge University Press, Cambridge.

Rescher, Nicholas
1978 *Scientific Progress: A Philosophical Essay on the Economics of Research in Natural Science.* University of Pittsburgh Press, Pittsburgh.
1980 *Unpopular Essays on Technological Progress.* University of Pittsburgh Press, Pittsburgh.
Rostow, W. W.
1980 *Why the Poor Get Richer and the Rich Slow Down.* University of Texas Press, Austin.
Shannon, Claude E.
1949 The Mathematical Theory of Communication. In *The Mathematical Theory of Communication*, by Claude E. Shannon and Warren Weaver, pp. 29–125. University of Illinois Press, Urbana.
Spengler, Oswald
1962 *The Decline of the West* (translated by Charles Francis Atkinson). Modern Library, New York.
Steward, Julian H.
1955 *Theory of Culture Change.* University of Illinois Press, Urbana.
Tainter, Joseph A.
1979 The Mountainair Lithic Scatters: Settlement Patterns and Significance Evaluation of Low Density Surface Sites. *Journal of Field Archaeology* 6:463–469.
1983 Settlement Behavior and the Archaeological Record: Concepts for the Definition of "Archaeological Site." *Contract Abstracts and CRM Archeology* 3:130–133.
1988 *The Collapse of Complex Societies.* Cambridge University Press, Cambridge.
1992 Evolutionary Consequences of War. In *Effects of War on Society*, edited by G. Ausenda, pp. 103–130. Center for Interdisciplinary Research on Social Stress, San Marino.
1994a Southwestern Contributions to the Understanding of Core-Periphery Relations. In *Understanding Complexity in the Prehistoric Southwest*, edited by George J. Gumerman and Murray Gell-Mann, pp. 25–36. Santa Fe Institute Studies in the Sciences of Complexity, Proceedings Volume XVI. Addison-Wesley, Reading.
1994b La Fine Dell'Amministrazione Centrale: Il Collaso Dell'Impero Romano in Occidente. In *Storia d'Europa, Volume Secondo: Preistoria e Antichità*, edited by Jean Guilaine and Salvatore Settis, pp. 1207–1255. Einaudi, Turin.
1995a Sustainability of Complex Societies. *Futures* 27:377–408.
1995b Valuing Complexity. In *Debating Complexity: Proceedings of the 26th Annual Chacmool Conference*, edited by Daniel Meyer, Peter Dawson, and Don Hanna. University of Calgary Archaeological Association, Calgary (in press).

Tainter, Joseph A., and G. John Lucas

1983 Epistemology of the Significance Concept. *American Antiquity* 48:707–719.

Tainter, Joseph A., and Fred Plog

1994 Strong and Weak Patterning in Southwestern Prehistory: The Formation of Puebloan Archaeology. In *Themes in Southwest Prehistory*, edited by George J. Gumerman, pp. 165–181. School of American Research Press, Santa Fe.

Weaver, Warren

1949 Recent Contributions to the Mathematical Theory of Communication. In *The Mathematical Theory of Communication*, by Claude E. Shannon and Warren Weaver, pp. 1–28. University of Illinois Press, Urbana.

White, Leslie A.

1949 *The Science of Culture.* Farrar, Straus and Giroux, New York.

1959 *The Evolution of Culture.* McGraw-Hill, New York.

Winterhalder, Bruce

1990 Open Field, Common Pot: Harvest Variability and Risk Avoidance in Agricultural and Foraging Societies. In *Risk and Uncertainty in Tribal and Peasant Economies*, edited by Elizabeth Cashdan, pp. 67–87. Westview, Boulder.

Jeffrey S. Dean
Laboratory of Tree-Ring Research, The University of Arizona, Tucson, Arizona 85721

Demography, Environment, and Subsistence Stress

INTRODUCTION

This paper is an attempt to specify aspects of a general framework for conceptualizing and analyzing the effects of subsistence stress and economic uncertainty on societies with subsistence-level agricultural economies. This effort is grounded in research on behavioral adaptation and paleoenvironment in the northern Southwest (Gumerman 1988) and in seminars on Southwestern societies sponsored by the School of American Research (SAR) (Gumerman 1994) and the Santa Fe Institute (SFI) (Gummerman and Gell-Mann 1994). The former produced a provisional model of behavioral adaptation to environmental variability that attempts to (1) develop a coherent conceptual framework for integrating environmental variability, human behavior, and human demography; (2) identify relevant environmental, behavioral, and demographic variables; (3) characterize these variables in ways that illuminate their relationships; and (4) specify potential adaptive interactions among the variables. The SAR seminar developed a regional and topical background for this paper and accumulated quantities of relevant demographic and behavioral data. The SFI workshop advanced several concepts of great potential explanatory power in the study of cultural evolution in general and Southwestern prehistory in particular. Among these ideas are the concept of cultures as complex adaptive systems, the relevance of general historical processes to the study of adaptive change (Wills et al. 1994), self-organization of complex systems (Kauffman 1991), poised states (Bak et al. 1988; Kauffman and Johnson 1992), evolutionary avalanches (Bak et al. 1988; Kauffman and Johnson 1992), and cultural evolution on rugged and coupled fitness landscapes (Kauffman and Johnson 1992; Kauffman and Levin 1987; Kauffman and Weinberger 1989). Determining the specific relevance of these concepts to the study of culture change and evolution and operationalizing them for this purpose will require much thought and directed research. In the current absence of this background, these concepts provide illuminating metaphors for objectifying conditions and processes that cause and direct culture change and evolution. Prehistoric behavioral adaptation in the Southwest provides an ideal case study for applying these ideas in a cultural context.

The degree to which the above ideas can further understanding of culture change and evolution is illuminated by the degree to which Southwestern adaptive behavior conforms to the processual principles that regulate the evolution of complex adaptive systems. Langton (Wills et al. 1994) identifies seven such principles that can be restated to apply to cultural systems as follows.

1. The processes involve populations of interacting individuals and social units.
2. Populations (or societies) behave as units, particularly vis-à-vis one another.
3. Higher levels of complexity emerge spontaneously, and often abruptly, through the specialization of individuals and social units.
4. Leaps to higher levels of complexity are characterized by greatly increased diversity, which subsequently declines as the best adaptive options are selected for.
5. Innovations diffuse within and among populations (societies).
6. Evolutionary processes operate on rugged or coupled fitness landscapes.
7. The operation of these historical processes results in evolution.

The perspective embodied in these principles has several implications for understanding the role of adaptive behavior in cultural evolution. First, as in any human cultural situation, we are dealing with complex adaptive systems operating within the context of general historical processes. Second, we are dealing with open systems in which dynamic fitness landscapes are transformed by both internal and external factors. Third, internal systemic relationships continually alter the fitness landscape as actors (individuals, societies, populations) adjust to new situations and as relationships change to accommodate these adjustments. Among these processes are autocatalytic changes in the fitness of the actors on the landscape and the alteration of the fitness landscape itself by changing behavioral relationships. Fourth, external variables change fitness landscapes through interchange among the actors, demographic factors, and environmental fluctuations.

ADAPTATION AND EVOLUTION

A discourse on adaptation and evolution took place at the workshop, stimulated by the presentation by Robert Leonard.[1] Having previously advanced a model of Anasazi behavioral "adaptation" (Dean 1988a), I wish to address a few salient points relating to "adaptation" and its role in sociocultural evolution. I use adaptation to refer to the integration of behavioral, demographic, and environmental variability to help maintain social entities at existing levels of complexity. Relevant entities include individuals, households, lineages, clans, moieties, villages, communities, tribes, on up to nation-states. Defined as a mechanism by which social units are perpetuated and transformed, behavioral adaptation is an important aspect of sociocultural change and evolution, although not all behavior is adaptive and not all adaptive behavior equals evolution.

Adaptation is an evolutionary outcome of selection in a Darwinian sense, although the adjective "natural" is not appropriate in this context. Because sociocultural elements are more important than biological ones in cultural situations, the unmodified term "selection" is used. A major unresolved issue in all applications of selection to human behavioral evolution is on what phenomena selection operates to produce evolutionary change. Numerous potential operands exist: populations, societies, cultures, beliefs, ideologies, organizational configurations, behaviors, artifacts, and others. In actual practice, selection can operate on any one or any combination of these factors, which makes cultural evolution an extremely complex process. For purely practical reasons, I focus on behavior as the target of selective agents in the evolution of cultural systems. Behavior subsumes a wide range of other factors including artifacts, activities, subsistence practices, social usages, and intergroup interactions. Furthermore, behavior has the advantage over many less concrete aspects of culture in being archaeologically perceivable. Whatever the debate over the study of prehistoric belief systems, ideologies, and cultural norms, there can be little argument that the archaeological record is a direct outcome of the actual behavior of human beings and can profitably be studied as such. Therefore, I focus on the selection of behaviors that enhance the survivability of particular groups of people, social units, life styles, belief systems, and, ultimately, cultures. Behavior is, of course, subsumed within and an expression of culture.

Selection, whether natural or cultural, operates on reservoirs of traits that, individually or in groups, increase or decrease in frequency as they are differentially favored or suppressed by the selective process. Within the context of a behavioral approach to cultural evolution, the relevant traits are behaviors that are selected for or against depending on their contribution to the survival of groups or cultural systems. Such behaviors need not be overtly expressed to be affected; they can be latent in the knowledge system of the group to be activated as situations require.

[1]The paper by Robert D. Leonard and Alysia L. Abbot, "Theoretical Aspects of Subsistence Stress and Cultural Evolution," is available from the authors (ed.).

Innovation (the invention of new traits or the recombination of existing traits) and the borrowing of traits from other groups increases the variability in the reservoir of potentially selectable behaviors. The broader and more variable the reservoir, the greater the supply of potentially selectable traits, which obviously enhances survivability.

Important aspects of evolution are the ways in which the traits originate and are transmitted among the units that comprise adaptive systems. New biological traits arise through mutation or other natural processes, while cultural traits can be borrowed from other groups, discovered accidentally or purposefully, created by altering relationships among existing traits, or invented in response to perceived needs. Biological transmission is effected by the genetic inheritance of traits from preceding generations. In contrast to the Mendelian inheritance of biology, the cultural transmission of traits (behaviors) is Lamarckian in the sense that acquired traits can be transferred among various entities. Furthermore, transmission is not restricted to direct biological or cultural descendants of the possessors of the traits. Rather, cultural elements circulate among the individuals and social units that make up a particular society and between different societies. Finally, unactivated behavioral elements can be preserved as knowledge (traditions) to be mobilized when needed. These aspects of element origin and transmission endow cultural systems with a wide range of variability on which selection can operate, and account in large measure for the unprecedented adaptability of such systems.

MODELING SOUTHWESTERN CULTURAL ADAPTATION

Elsewhere (Dean 1988a), I present a preliminary general model of behavioral adaptation for Anasazi populations in the northern Southwest. The model is an abstract construct designed to identify variables that operate in behavioral adaptational situations on the Colorado Plateau and to specify dynamic relationships among these variables. The model, as formulated, is not intended to be operationalized through the control and measurement of individual variables. Operationalization is left to studies of specific instances of adaptive behavioral change in which relevant variables can be defined, isolated, and measured. The conceptual model was developed as a formal mechanism for integrating various kinds of paleoenvironmental and archaeological data into the study of human behavioral adaptation and evolution on the Colorado Plateau. This approach was taken in the conviction that unless a conceptual scheme existed to relate environmental variability to human behavior, the effort to identify prehistoric environmental impacts on Southwestern groups would degenerate into an inconclusive exercise in pattern matching. Because this model has been published (Dean 1988a; Dean et al. 1985; Plog et al. 1988), I only summarize and amplify salient points pertinent to resource stress and economic uncertainty.

Three basic premises underlie the model. First, for a population, group, or society to survive and perpetuate itself, its members must be able to make a satisfactory living in the environment in which it exists. Second, since the environment varies through time and space at rates too high to be accommodated by genetic adaptation, societies (cultures) must possess the behavioral flexibility to accommodate these changes. Third, sociocultural evolution occurs when selection causes certain behaviors to be emphasized over others and when the integration of the resulting changes into behavioral systems creates new adaptive configurations.

Within the context of the model, adaptation is viewed as proceeding through the process of selection. Adaptation is the outcome of both selection for behaviors (and other traits) that enhance a society's fitness and selection against behaviors that reduce fitness. Contrary to some misunderstandings of the model, adaptation is not viewed as a process involving conscious adjustments to environmental fluctuations, although such efforts may expand the reservoir of behaviors on which selection operates. Thus, the concept of adaptation embodied in the model is compatible with a "selectionist" approach to sociocultural evolution (Dunnell 1980; Leonard 1989; Leonard and Jones 1987).

Adaptation results from the interaction of three major classes of variable (Kirch 1980): environmental, demographic, and behavioral (Dean 1984, 1988a; Dean et al. 1985). In order to relate these factors to human adaptation, the model arbitrarily partitions the continua of variability within each class into three types—stable, low frequency, and high frequency—on the basis of temporal structure. Stable factors have not changed appreciably during the time frame of the study; for the purposes of this paper, the last 3,000 years. Because these factors have not changed, their present states accurately reflect past conditions. Low- and high-frequency factors exhibit primary periodicities that, respectively, are greater or less than 25 years, the length of a human generation. The relevant periodicities are those of the natural processes that control the various factors, and even low-frequency processes can cause abrupt environmental changes. Because low- and high-frequency factors vary within the study interval, they must be reconstructed by relevant paleoenvironmental techniques.

Stable environmental factors that have changed insignificantly during the last 3,000 years include bedrock geology, regional hydrology, gross topography, vegetation zonation, and climate type. Low-frequency environmental variables with base periodicities greater than or equal to 25 years include the rise and fall of alluvial groundwater levels and the deposition and erosion of floodplain sediments along drainages (Karlstrom 1988), changes in effective moisture and in the composition and elevational boundaries of plant communities (Hevly 1988), and long-term trends in climate. High-frequency environmental variables are primarily climatic and include precipitation and temperature variability on scales ranging from days to decades (Dean 1988b).

Demographic variables include population size, birth and mortality rates, sex composition, age structure, and health. The first two of these have been the focus of most studies of prehistoric population variation (Powell 1988). Sheer numbers of people, of course, have major implications for a society's survival. Many effects of population size variability are, however, related to density rather than just magnitude. Thus, certain population densities are necessary to support specific organizational configurations or to permit or preclude certain adaptive behaviors. Conversely, population decimation has numerous important and well-known social effects. While sex and age structure and health profiles undoubtedly have crucial adaptive consequences, they are inadequately represented archaeologically and contribute little to general studies of prehistoric cultural evolution. Like environmental factors, demographic variables exhibit a range of variation, although long-term stability rarely has been achieved, at least in the Southwest (Dean et al. 1994). Population size variability has both low- and high-frequency components. Most population curves exhibit long-term trends with periods greater than 25 years as well as more rapid fluctuations that fall within the high-frequency range.

Inherently variable, human behavior exhibits few stable characteristics but does have both low- and high-frequency components. Low-frequency processes impart great inertia to cultural systems, which accounts for the persistence of general patterns such as the Puebloan lifeway. Cultural traditions exert a powerful effect on behavioral adaptation by contributing functionally and ideologically coherent sets of behaviors to the reservoir of traits that are acted on by selection. Internal linkages among these elements account for the transmission and selection of assemblages of traits rather than single behaviors. High-frequency processes are responsible for behavioral variation within sociocultural configurations perpetuated by low-frequency processes. These processes (discovery, innovation, borrowing, modification, and integration) produce much of the behavioral variability that endows cultural systems with their noted selective flexibility and unparalleled evolutionary potential.

Selection of alternative behaviors is not unlimited; rather, it is constrained by conditions and factors that characterize particular adaptive situations. Existing cultural configurations (social structure and organization, religious systems, technology, ideology, etc.) influence the range of acceptable behavioral elements and the ways in which these elements can be altered, engaged, and integrated into existing adaptive situations. Thus, newly acquired or emphasized behaviors usually are altered to make them compatible with existing cultural patterns. In addition, environmental and demographic factors set boundaries on the effectiveness of particular technological and social adaptive mechanisms. For example, different environments provide different potentials for subsistence behavior, while certain population densities are necessary to support more complex social organizations. Finally, numerous historical factors and events (such as interactions with other groups, warfare, conquest, and colonization) limit appropriate behavioral options.

OPERATION OF THE MODEL

Relationships among environmental, demographic, and behavioral variables are specified by a particular concept of carrying capacity: the maximum number of people that can be supported by a given subsistence technology under prevailing environmental conditions (Hassan 1978:73). Carrying capacity is viewed not as a static, measurable limit but as a dynamic boundary that fluctuates with changing environmental, population, and cultural conditions. Conceived as a dynamic phenomenon, carrying capacity corresponds in some ways to the concept of a coupled fitness landscape (Kauffman and Johnson 1992) on which variables of different types interact in response to selection and systemic processes. Adaptively effective changes in the components of any one of the variable classes alter carrying capacity relationships and deform fitness landscapes into new configurations with different selective potentials. Within this conceptual framework, resource stress and/or economic uncertainty develop when population numbers approach or exceed carrying capacity thresholds. When resource stresses occur, adjustments to preserve the existing population take place. Since societies of the level of complexity of those in the prehistoric Southwest have limited technological capacity purposefully to improve the natural environment, many of these adjustments involve the demographic and behavioral components of the adaptive situation.[2]

Figure 1 illustrates ways in which carrying capacity thresholds can be breached and some possible responses to these transgressions. The heavy, sinusoidal lines and superimposed lighter lines represent, respectively, the basic carrying capacity established by environmental fluctuations due to low-frequency natural processes and high-frequency oscillations around the fundamental threshold. The medium-weight lines represent low-frequency trends in population with high-frequency variations indicated by superimposed lighter lines. The rectangles identify "zones of interference" in which high-frequency fluctuations in population and environmental variability intermittently overlap. Heavy dashed lines indicate changes in carrying capacity effected by human behavioral reaction to resource stress.

Adaptive crises develop when population numbers approach or surpass the carrying capacity (Figure 1(a)) or when environmental degradation lowers carrying capacity below the existing population level (Figure 1(b)). In either case, high-frequency environmental and demographic fluctuations converge before population actually intersects the base carrying capacity. Such "interference" signals an impending adaptational crisis and could trigger cultural reaction before the fundamental carrying capacity actually is breached. If such early warnings fail to provoke a response, adaptive behavioral (Figure 1(a)–1(b)) or demographic (Figure 1(c)) transformations would follow transgression of the basic threshold. The former involve behavioral processes that increase the carrying capacity, while the

[2] Compare the chapter by Alan Sullivan (this volume).

latter include processes that maintain population levels around or below the carrying capacity.

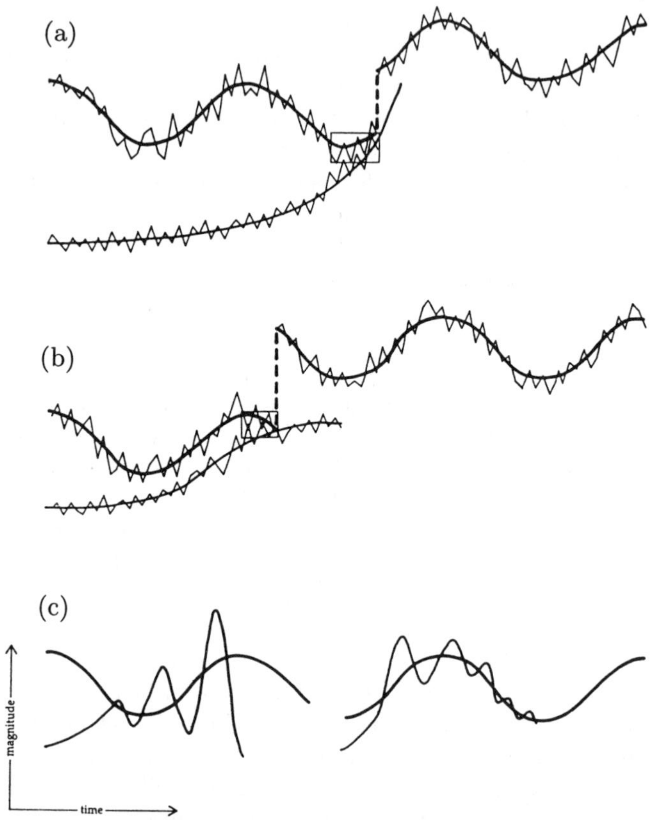

FIGURE 1 Interaction of environmental, demographic, and behavioral variability in the Southwestern adaptation model. (a) Growing population breaches the carrying capacity threshold determined by low-frequency environmental processes. (b) Environmental deterioration lowers the carrying capacity limit below the level of the population. (c) Deviation-amplifying (left) and deviation-reducing (right) population responses to transgressions of carrying capacity boundaries. The heavy sinusoidal lines represent fluctuations in basic carrying capacity caused by low-frequency environmental processes; the superimposed light lines represent high-frequency environmental oscillations; the medium and superimposed light lines indicate, respectively, low- and high-frequency variability in population numbers; the dashed lines represent behavior-induced increases in carrying capacity; the rectangles specify "zones of interference" between high-frequency environmental and population variability.

Population adjustments to resource stress can be either purposeful or imposed. Intentional responses include the emigration of part of the population into nearby underpopulated areas, the regulation of birth rates, or the activation of extreme measures such as infanticide. Any of these actions could produce the deviation-amplifying profile (Figure 1(c)), when successive adjustments miss the carrying capacity by increasingly large margins (Plog 1986), or a deviation-reducing profile (Figure 1(c)), when adjustments more closely approximate the limit. Should such purposeful demographic responses fail to remedy the situation, involuntary population adjustments due to increased mortality rates, decreased birth rates, malnutrition, or, in extreme cases, starvation are likely.

Behavioral responses to resource stress occasioned by the breaching of carrying capacity boundaries generally increase the carrying capacity by altering the cultural component of the adaptive situation. These adjustments include (1) technological fixes that increase production either through the application of more productive subsistence methods or through changes in the amount and/or organization of labor invested in the production system (Boserup 1965), or (2) sociocultural changes that improve the acquisition, accumulation, preservation, and distribution of resources. Technological responses include things such as monocropping, increased irrigation, and increased individual labor devoted to farming. Sociocultural responses include things such as changes in the organization of food production and storage, increased centralization of food accumulation and distribution, and expanded trade relationships. Potentially ameliorative behavioral traits are extracted from the reservoir of variability already extant in the culture, are borrowed from other groups, or, on occasion, are invented in response to the perceived need (Barnett 1953). Through the process of selection, behaviors that increase the fitness of the group become more frequent relative to others that are less efficacious, thereby creating new behavioral adaptations that are fitter than their predecessors. These changes precipitate numerous internal adjustments in the systemic relationships among individuals and social units and ultimately create new cultural configurations that are qualitatively and quantitatively different from their predecessors.

Contrary to some arguments that an adaptive approach to culture change is incompatible with a selectionist approach to cultural evolution (Leonard 1989; Leonard and Jones 1987), the model outlined here is congruent with selectionist theory because, in both formulations, adaptation is viewed as an outcome of selection, be it natural or cultural. This congruity is demonstrated by a quasi-hypothetical example from Long House Valley in northeastern Arizona. Dean and Lindsay (1978) and Dean et al. (1975, 1978) attribute a major change in settlement pattern and social organization that occurred between A.D. 1150 and 1250 to behavioral adaptation to major resource stress caused by low- and high-frequency environmental changes. Figure 2(a) illustrates a serious environmental degradation that, around 1150, triggered important changes in agricultural practice that were necessary for the population to maintain itself in the valley. These subsistence changes transformed the pre-1150 settlement pattern of regular site distribution around the margin of the floodplain (State I in Figure 2(b)) to the post-1250 pattern of site clusters located in areas of the valley floor that were farmable under the post-1150 environmental conditions (State II in Figure 2(b)). This economic

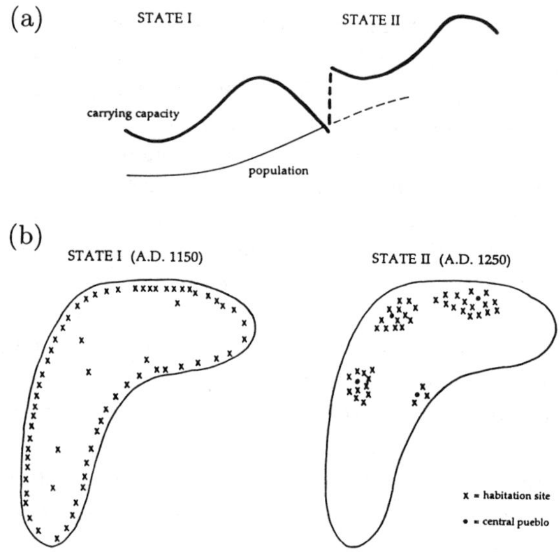

(c)

AGRICULTURAL PRACTICES	STATE I	STATE II
groundwater farming	70%	25%
Ak Chin farming	1%	25%
sand dune farming	2%	15%
upland dry farming	15%	1%
terracing	9%	10%
irrigation	2%	15%
spring farming	1%	5%
arroyo bottom farming	0%	3%
pot watering	0%	1%
SETTLEMENT BEHAVIOR		
location	anywhere	restricted farming loci
pattern	evenly dispersed	clustered
hierarchy	one level	four levels
maximum site size	15 rooms	400 rooms
integrative architecture	hamlet kivas	kivas, plazas, spinal roomblocks

FIGURE 2 Schematic example of a selectionist perspective on a prehistoric adaptive transformation in Long House Valley, northeastern Arizona. (a) Behavior-induced increase in carrying capacity in response to population-resource imbalances caused by environmental deterioration around A.D. 1150. (b) Settlement changes between 1150 (State I) and 1250 (State II) caused by adaptive responses to resource stress (x = habitation site, • = central pueblo). (c) (1) Hypothetical percentage changes in agricultural systems caused by the differential selection of practices more suited to the State II situation, and (2) adaptive responses in settlement caused by changed ecological relationships and social adjustments to aggregation.

transformation was accomplished in part by the differential selection of various agricultural methods so that the State II mix of techniques differed significantly from that of the State I mix (Figure 2(c)). Note that the percentages are hypothetical and serve only to exemplify the way in which selection operates to create a new adaptation to altered circumstances. The redistribution of settlements greatly increased local population densities and created numerous social problems that had not existed when settlements were evenly dispersed. Major organizational changes that materially increased the level of social complexity were selected for to resolve these problems. Thus, environmentally induced changes in the adaptive situation triggered selection processes that, in turn, caused evolutionary changes in the local culture.

PREHISTORIC SOUTHWESTERN DEMOGRAPHY AND ENVIRONMENT

The conceptual model identifies low- and high-frequency fluctuations in demography and the environment as important independent variables and behavior as the dependent variable in most adaptive situations. Obviously, if our goal is the study of resource stress, we wish to isolate instances with high potential for the intersection of the population and environmental curves on either local or regional scales of analysis. Low- and high-frequency variations in both human population and the environment must be accurately reconstructed before these factors can be used to identify circumstances in which selective forces operate to produce adaptational change. Estimating population numbers is one of the most daunting tasks that faces archaeology (Powell 1988). For my purposes here, I rely on a synthesis of Southwestern population estimates produced by Dean et al. (1994) as part of an investigation of the organization and evolution of Southwestern societies undertaken by the School of American Research and the Santa Fe Institute. Low-frequency environmental reconstructions are based on alluvial chronostratigraphic, palynological, packrat midden, and some dendrochronological studies, while high-frequency environmental variability is reconstructed primarily through dendroclimatic analyses and some pollen studies (Gumerman 1988).

Resource stress and economic uncertainty can be studied on regional or local scales. The latter is likely to be the most productive approach because even large-scale environmental and population fluctuations have specific local expressions and consequences and because the scale of Southwestern societies precluded coordinated regional-level responses. Nevertheless, regional demographic and environmental characteristics are more appropriate to the general consideration presented here.

DEMOGRAPHY

The regional population trend (Dean et al. 1994:Figure 12) resembles a normal curve that is skewed somewhat toward the early end, that peaks at 100,000+ individuals between A.D. 1000 and 1250, and that tails off to approximately 10,000 people at A.D. 1600. An apparent growth spurt between A.D. 550 and 650 is due more to the greater archaeological visibility of Basketmaker III over earlier manifestations than to a real demographic florescence. A steeper increase after A.D. 800 is due primarily to growth in the Sonoran Desert and the San Juan Basin. A minor downturn in regional population after A.D. 1000 reflects population decrease in the San Juan Basin. A steep decline afterA.D. 1250 is partly real and partly due to the paucity of archaeological research in areas inhabited after A.D. 1300. Because the regional population curve is the sum of many local population estimates, it lacks the high variability that characterizes most local sequences. The curve does indicate that population probably did not approach the regional carrying capacity before A.D. 900, although local encroachments of fluctuating carrying capacity limits undoubtedly occurred earlier.

The least variable "local" population curve is that for the Sonoran Desert (Dean et al. 1994:Figure 11), which rises fairly rapidly from A.D. 600 to a relatively stable peak of around 25,000 people that lasts from about A.D. 1000 to 1250 and falls off after 1250. This apparent stability is due to two factors. First, the Hohokam curve is a fairly large-scale reconstruction that masks the variability in its local constituents. Second, the curve reflects the population stability afforded by irrigation from large, through-flowing rivers supplemented by the vegetational bounty of the Sonoran Desert. Hohokam population trends provided a fairly stable floor for the regional population.

Populations of the Mogollon Highlands and localities on the Colorado Plateau are much more variable than those of either the Desert or the Southwest as a whole. The Mogollon curve (Dean et al. 1994:Figure 10) rises steadily to a peak of around 4000 people in the middle A.D. 900s, then exhibits a series of increasingly large fluctuations that resemble the classic deviation-amplifying pattern (Plog 1986), and falls to zero between A.D. 1350 and 1450. Because they represent fairly small areas, most Colorado Plateau population curves (Dean et al. 1994:Figures 1–9) exhibit considerable variation. Two peripheral areas (the Virgin Branch and the Grand Canyon) and the Mimbres area have unimodal curves that peak in the middle 1100s, while the northern Rio Grande area has a unimodal curve that peaks around 1300. Interior Colorado Plateau areas (the Mesa Verde area, the Kayenta heartland, the Little Colorado River drainage, the San Juan Basin, and the Cebolleta Mesa area) are multimodal with maxima in the eleventh (San Juan Basin), thirteenth (Mesa Verde, Kayenta, Cebolleta), and fourteenth (Little Colorado) centuries. These fluctuations created numerous opportunities for local population-resource imbalances (Cordell and Plog 1979) that would have created conditions of resource stress and economic uncertainty.

ENVIRONMENT

Several aspects of the Southwestern environment fall into the stable category. The familiar tripartite division of the region into plateau, mountains, and desert is a long-standing feature based on other stable factors such as general climate, elevation, aspect, geology, topography, and vegetation. A single climate type has characterized the region since the end of the Pleistocene (Schoenwetter 1962), even though considerable spatial and temporal variability exists within the limits of the prevailing climatic regime.

Regional-scale low-frequency variables are more difficult to identify because comparable paleoenvironmental reconstructions are not available for all areas. Alluvial chronostratigraphic studies are rare in the mountains and those in the desert lack the chronological controls of their northern counterparts. Nevertheless, geological studies in the desert (Eddy and Cooley 1983; Sayles 1983; Waters 1986) reveal alluvial sequences that are morphologically similar to and roughly contemporaneous with those of the Colorado Plateau (Dean 1987). Until other research in the Southwest contradicts these conclusions, it probably is justifiable to project the general configuration of the Colorado Plateau alluvial reconstructions into the mountains and desert. Palynological studies are more widespread, but differences in levels of chronological control and environmental resolution hinder correlation and comparison of local pollen sequences. As yet, a regional pollen reconstruction has not been synthesized.

Regional-scale high-frequency variability is even more difficult to isolate because long tree-ring chronologies, the principal basis for such reconstructions, are not available for the desert. Furthermore, most dendroclimatic reconstructions are, by nature, highly localized with little in the way of a regional signal. Attempts to derive such a signal by merging ring sequences from across the region reduces the variability in the chronologies and damps the climatic signal. Measures of temporal and spatial variability in dendroclimate are most likely to have large-scale implications, but even these are of unknown relevance to the desert.

Low- and high-frequency environmental variability is much better understood on subregional levels of analysis. The low-frequency hydrologic and aggradation-degradation curves (Figure 3A,B) represent, respectively, the rise and fall of alluvial groundwater levels and the deposition and erosion of floodplain sediments on the plateau (Karlstrom 1988). The effective moisture curve (Figure 3C), derived from pollen analyses (Hevly 1988), represent low-frequency fluctuations in moisture available to plants and conform remarkably well to the hydrologic curve. These variables are crucial to agricultural production, especially that derived from cultivation on floodplains (Dean 1988a). Rising or high water tables and aggrading or undissected floodplains create optimal conditions for groundwater farming that is relatively independent of precipitation. In contrast, falling or low water tables and degrading or dissected floodplains remove the groundwater from arable surfaces and destroy large areas of alluvial bottomland, thus making agriculture more

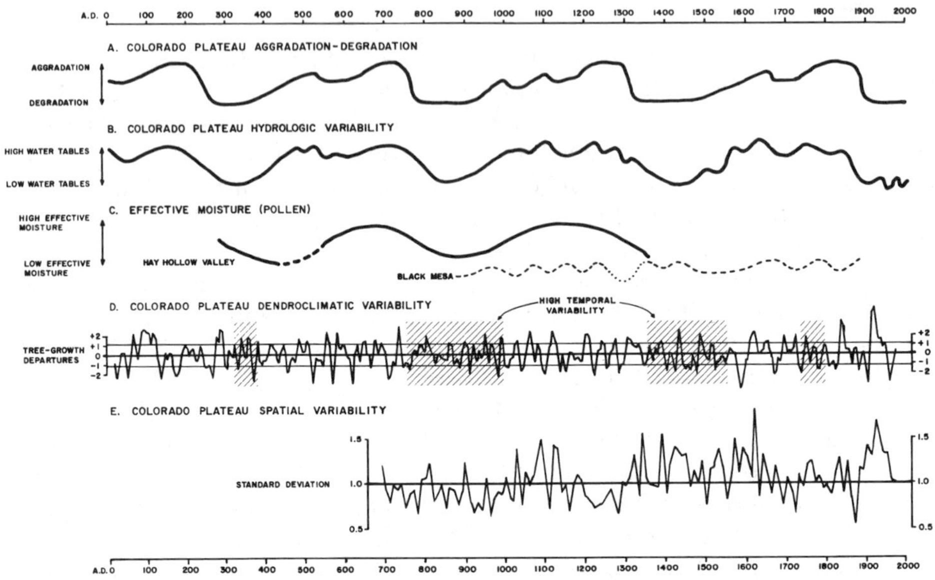

FIGURE 3 Low- and high-frequency environmental variability in the northern Southwest. A. Primary and secondary variations in alluvial groundwater levels. B. Deposition and erosion of alluvium. C. Palynologically determined fluctuations in effective moisture. D. Decadal variability in relative dendroclimate (hatching indicates periods of high temporal variability). E. Spatial variability in dendroclimate.

dependent on a less sufficient and more variable source of water, precipitation. Environmental deterioration of this type creates major resource stress for subsistence systems based on floodplain farming, such as those of the eastern Kayenta area, but has less impact in areas, such as those around Mesa Verde and Navajo Mountain, where upland dry farming predominated. In upland areas, high-frequency precipitation variability would have been more limiting; however, the alluvial curves may reflect other low-frequency processes that affected upland agricultural systems. In either situation, low-frequency processes affect the relative emphasis on upland vs. lowland agriculture (Plog et al. 1988).

High-frequency environmental variability is reconstructed through dendroclimatic analyses of a network of 27 climate-sensitive archeological tree-ring chronologies that covers the Southwest north of the Gila River (Dean and Robinson 1978). Using these data, Dean and Robinson (1977) reconstructed relative variations in annual precipitation for the northern Southwest, and Graybill (1989) reconstructed

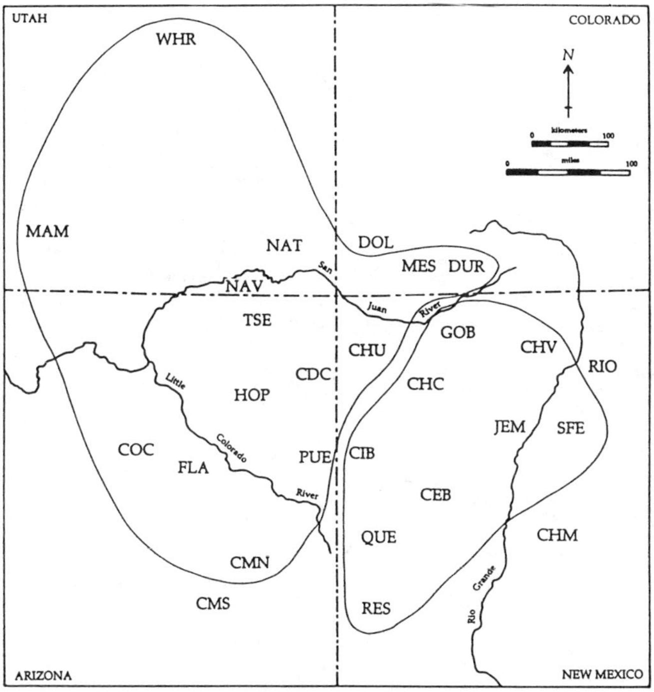

FIGURE 4 Principal components of Southwestern tree growth (climate) for the period A.D. 966–1988.

streamflow for the Salt and Verde Rivers. A recent large-scale reconstruction project by Dean, Graybill, and Funkhouser has produced quantitative reconstructions of precipitation and Palmer Drought Severity Indices for each of the 27 stations, augmented the Salt-Verde and Gila Rivers streamflow reconstructions, and developed several measures of climatic variability across the station grid. Figure 3D illustrates amplitude and temporal variability in dendroclimate across the northern Southwest. Amplitudes are expressed as decadal positive and negative departures (in standard deviation units) from the long-term chronology mean and indicate periods of above- and below-average precipitation. The time series is characterized by alternating periods of high temporal variability (hatched), when minima and maxima alternated rapidly, and low temporal variability (unhatched), when the transitions from extreme high and low values were more gradual. The spatial

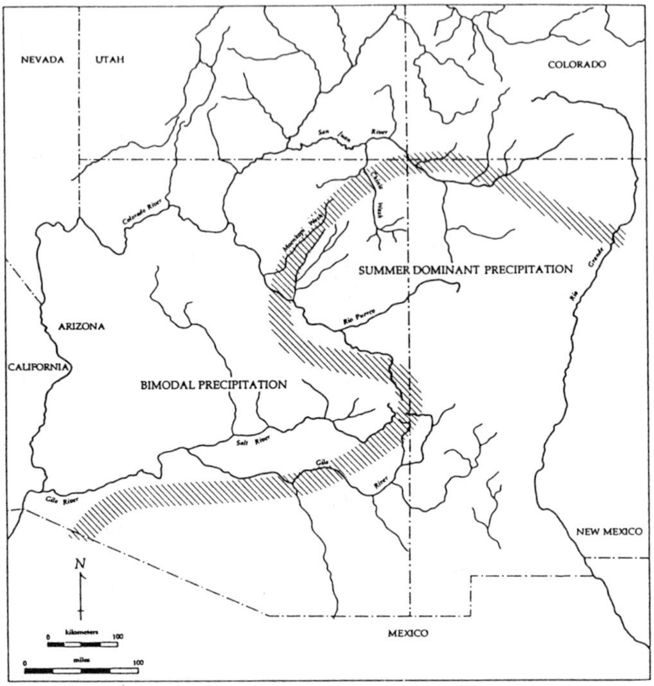

FIGURE 5 Spatial distribution of seasonal precipitation in the Southwest during the last century.

variability curve (Figure 3E) indicates whether climatic conditions differed among areas (high variability) or were similar across the region (low variability).

Dean, Graybill, and Funkhouser used principal components analysis (PCA) of the 27 tree-ring chronologies to further characterize climatic variability across the region. Analysis of the grid for the common period A.D. 966 to 1988 segregated the stations into two significant principal components (Figure 4). PCA 1 loads heavily on stations in the northwestern two thirds of the Southwest, while PCA 2 loads most heavily on stations in the southeastern third of the region. This long-term regional pattern bears a provocative resemblance to the present distribution of annual precipitation in the region (Figure 5), which is characterized by a unimodal, summer-dominant regime on the southeast and a bimodal, summer-winter regime to the northwest, separated by a sinuous transition zone that winds through Arizona and New Mexico. The resemblance is strong enough to suggest that the spatial distribution of the two dominant principal components represent the persistence of the unimodal vs. bimodal precipitation pattern for at least the last millennium.

The two components exhibit systematic attribute differences that probably reflect the inferred differences in precipitation regime. The five-year means (Figure 6) for PCA 2 (southeast) have higher amplitudes than do those for PCA 1 (northwest) except after about A.D. 1820 when PCA 1 amplitudes slightly exceed their PCA 2 counterparts. Amplitudes seem to diminish fairly regularly in the southeast while remaining comparatively stable in the northwest until 1820, when they increase abruptly. Five-year mean standard deviations (Figure 7) also differ between the two components. The PCA 2 standard deviations exhibit little temporal trend, although they do increase in amplitude from early to late. In contrast, the PCA 1

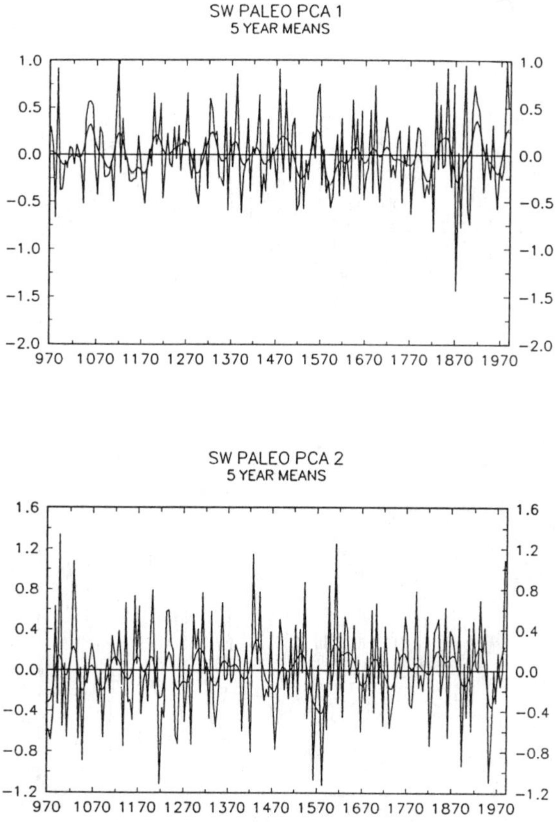

FIGURE 6 Five-year averages of mean ring indices for the two main principal components of Southwestern tree growth. PCA 1 (top) includes index chronologies in the northwestern two-thirds of the region; PCA 2 (bottom) includes stations in the southeastern third of the region.

standard deviations exhibit greater amplitudes prior to about A.D. 1300 and evince a strong positive trend after 1320. These differences are consistent with a stronger winter precipitation component in the northwest compared a summer-dominant regime in the southeast.

In order to examine the temporal structure of the long-term patterning, we did PCA analyses on the grid for successive 100-year intervals overlapped by 50 years from A.D. 689 to 1988. Nearly all the intervals produced patterns similar to the overall configuration (Figure 8). This outcome demonstrates that the observed southeast-northwest duality is a long-term aspect of Southwestern climate that has

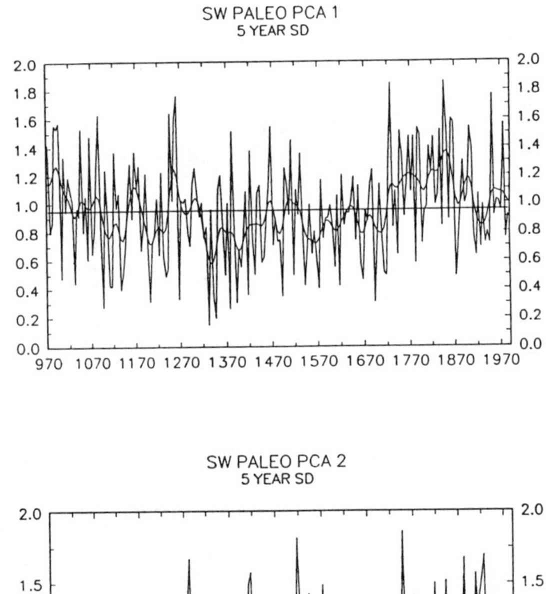

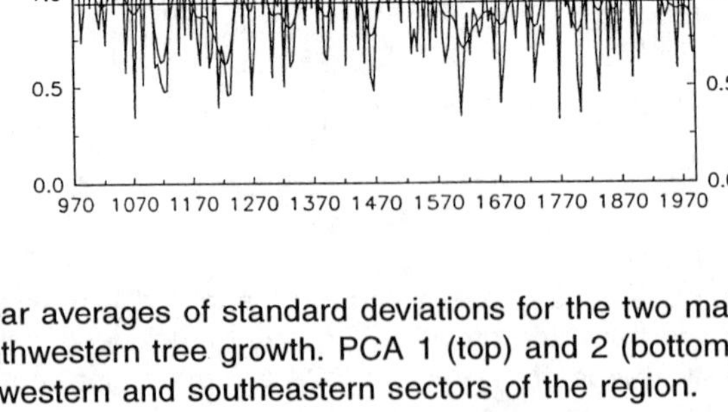

FIGURE 7 Five-year averages of standard deviations for the two main principal components of Southwestern tree growth. PCA 1 (top) and 2 (bottom), respectively, represent the northwestern and southeastern sectors of the region.

prevailed for the last 1500 years. The picture is quite different, however, for the period between about A.D. 1250 and 1450 when the long-term pattern broke down into chaotic distributions of three or four principal components that exhibit no logical geographic patterning. Figure 9, showing the A.D. 1339–1438 interval, exemplifies the configurations of this period. Obviously what occurred was a 200-year, regional-scale disruption of a climatic pattern that characterized the Southwest for the preceding 550 years and the following 550 years.

The exact climatic meaning of this disruption is not yet clear, but some inferences regarding its nature can be made. First, the persistence of the southeastern component (PCA 2 in other periods) through the A.D. 1250–1450 interval indicates that the unimodal, summer-dominant rainfall regime has been stable over the entire 1500-year period of record. Second, the changes between A.D. 1250 and 1450 are

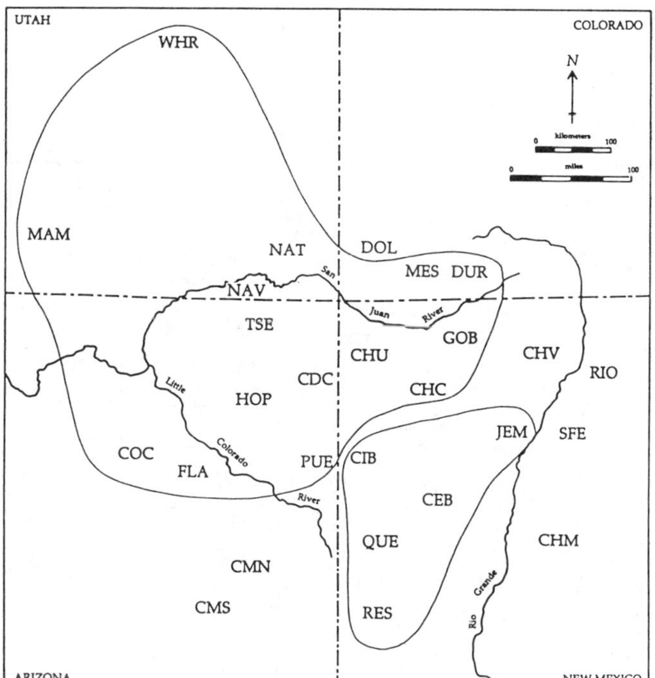

FIGURE 8 Principal components of Southwestern tree growth (climate) for the period A.D. 739–838.

concentrated in the northwestern area, which suggests that the bimodal precipitation pattern gave way to a variable mixture of rainfall regimes. Thus it appears that most of the disruption took place in the northwestern subregion and that the climatic situation remained relatively unaltered in the southeastern area.

REGIONAL RESOURCE STRESS AND ECONOMIC UNCERTAINTY

Within the context of the adaptation model, the population and environmental reconstructions summarized above can be used to identify instances of regional

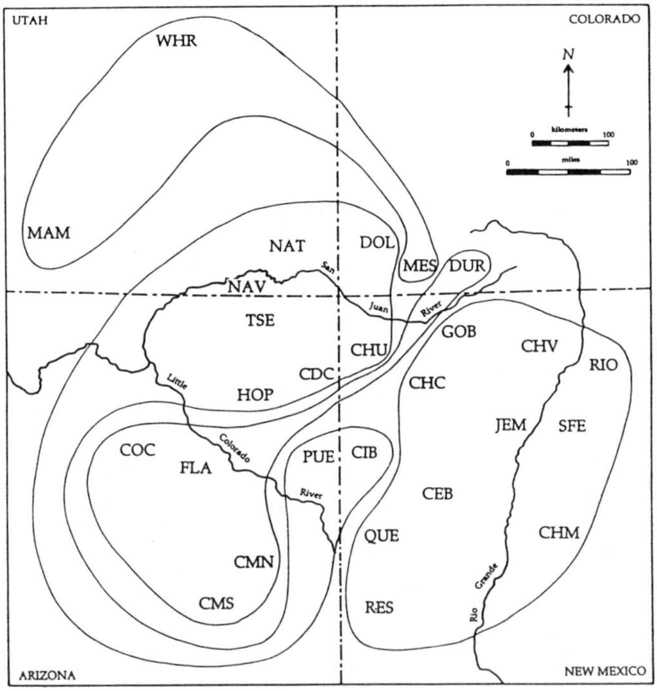

FIGURE 9 Principal components of Southwestern tree growth (climate) for the period A.D. 1339–1438.

resource stress serious enough to have triggered large-scale adaptive behavioral change. Although local populations undoubtedly breached carrying capacity limits on many occasions, large-scale adaptive crises of this kind probably did not occur until regional populations approached saturation levels, a point probably not achieved until the tenth or eleventh centuries. Furthermore, until regional population saturation was achieved, many local transgressions of carrying capacity boundaries could be met through mobility strategies that involved all or part of a group moving into sparsely occupied areas, which amounts to transferring an existing adaptive configuration to a more suitable location. After about A.D. 900, however, when most suitable areas were occupied by burgeoning populations, this option became less viable, and transformations of behavioral systems proliferated. As a result, regional-scale behavioral responses to resource stress and economic uncertainty should not be expected much before 900.

Because basic carrying capacity levels are established by stable and low-frequency environmental factors (Dean 1988a), variation in the latter can be used to pinpoint instances of potential adaptive stress. Thus falling and low alluvial water tables and active floodplain erosion coupled with high population levels should mark periods of greatest resource depletion, through loss of land and water, and therefore greatest resource stress, especially for agriculturalists. High-frequency variability can exacerbate or alleviate such stress when, for example, below- or above-average annual precipitation, low or high spatial variability, or negative or positive temporal trends accompany unfavorable low-frequency conditions. Similarly, these factors can diminish or enhance beneficial low-frequency conditions, especially when population approaches carrying capacity.

Given low regional population, the primary alluvial degradation between A.D. 200 and 400 is likely to have had few large-scale consequences as mobility allowed populations to alleviate local resource stresses. The high temporal climatic variability of that time also is unlikely to have had many large-scale effects. About the only major regional development of this period is the widespread appearance of agricultural hamlets composed of shallow pithouses and facilities for storing crops (Matson 1991; Smiley 1985; Wills 1988). Even with low population, deleterious alluvial conditions may have been partly responsible for this development by reducing arable acreage and forcing farmers to concentrate in localities where agriculture remained possible. The contraction of suitable farming area would have limited the mobility response and elevated local population densities. It is interesting that, on the Plateau at least, this "aggregated" settlement pattern virtually disappeared after A.D. 400 when improved low-frequency conditions eased the constraints on mobility.

Given markedly increased population, the primary alluvial degradation between A.D. 750 and 900 should have had more obvious regional consequences than the previous one. Suppressed climatic amplitudes would have mitigated low-frequency stress while high temporal and low spatial climatic variability would have intensified it. Although numerous local and even subregional developments, such as major changes in Kayenta Anasazi settlement pattern and community organization (Dean

1970, 1991) and the beginning of the Chacoan regional interaction system (Vivian 1990), occurred at this time, few major regional developments can be identified, perhaps because population had not yet reached a critical level.

Greatly increased population after A.D. 1000 should amplify the effects even of less-severe low-frequency declines, such as the secondary alluvial degradation in the middle 1100s. A major drought between 1130 and 1180 coupled with low temporal variability would have augmented the low-frequency stress. Declining spatial climatic variability would have limited regional exchange as a means of offsetting local production shortfalls. Resource stress caused by falling carrying capacity and increasing population undoubtedly was an important factor in the major cultural changes that took place in the middle twelfth century. The Chacoan system ceased expanding and former components of this interaction system began to diverge from the Chacoan pattern, widespread down-elevation settlement displacements caused the depopulation of numerous upland areas, populations began concentrating in low-elevation areas with reliable supplies of water, the peripheries of the Anasazi area were abandoned as populations contracted back into the core areas, puebloan community configurations replaced pithouse villages in the Mogollon Highlands, the Mimbres culture waned, and the transition from Sedentary to Classic occurred among the Hohokam. The new adaptational relationships created by these changes made the societies of the region even more vulnerable to population-resource imbalances.

Even greater resource stress developed after A.D. 1250 when elevated population densities caused by settlement responses to the mid-twelfth-century crisis coincided with serious low- and high-frequency environmental degradation. Falling alluvial water tables and floodplain erosion, severely depressed annual rainfall between A.D. 1275 and 1300 (the "Great Drought"), low temporal and spatial climatic variability, and the breakdown in the Southwestern precipitation pattern combined to lower regional and local carrying capacities at a time when populations had become increasingly aggregated. This interval was marked by far-reaching demographic and sociocultural changes that undoubtedly derive in part from major resource stresses and economic uncertainties occasioned by substantial population-resource imbalances. The San Juan drainage was virtually abandoned as a place of residence for agricultural societies as the population shifted south and east into the Little Colorado Valley, the Mogollon Highlands, and the Rio Grande drainage. The rise of extremely large communities in these areas was stimulated in part by conditions more favorable for agriculture under the altered environmental regime and in part by new socioreligious developments, such as the katsina cult (Adams 1991), that attracted immigrants. The Sonoran Desert saw major changes with the development of the Hohokam Civano Phase and the florescence of the Salado pattern.

The exact role of demographic and environmental variability and of resource stress in the major cultural transformations that occurred after A.D. 1250 remain uncertain and require additional, focused research. There can be little doubt, however, that high population densities coupled with both low- and high-frequency environmental deterioration would have created major population-resource imbalances that would have to be accommodated for the extant societies to survive. The contribution of the disruption in the long-term regional precipitation pattern revealed by the principal components analysis to the resource stress of the late thirteenth century likewise remains problematical. It is significant, however, that the area most severely impacted by this discontinuity, the northwestern sector, is the homeland of Anasazi populations who could not have been immune to this climatic excursion. The societies of this subregion undoubtedly were well adapted to the bimodal precipitation regime that had persisted for at least 550 years. The interruption of this strong pattern undoubtedly would have severely altered established adaptive configurations, especially given the relatively high populations of the period. That the numerous cultural changes between A.D. 1250 and 1450, which included the virtual abandonment of the northwestern sector and large-scale movement into the climatically more stable southeastern subregion, are in some way associated with the disintegration of the regional climatic pattern seems obvious. Considerable climatic, dendroclimatic, and archaeological research remains to be done in order to specify the dynamics of this situation.

The reassertion of the "normal" climatic pattern after A.D. 1450 coincided with the resumption of groundwater accretion and alluvial deposition, reduced climatic amplitudes, high temporal climatic variability, increasing spatial variability, lower population levels, and the development of more efficient and sophisticated agricultural systems. All these factors would have combined to alleviate population-resource imbalances, reduce resource stress, and increase economic stability. The return to sociocultural stability after 1450, albeit in different areas and highly altered environmental and cultural circumstances, probably reflects at least in part the ameliorating adaptive situation.

Before the next major low-frequency environmental degradation occurred near the end of the nineteenth century, the imposition of European colonialism through conquest transformed the adaptive situation by changing intersocietal relationships and by upsetting relationships between humans and the natural environment that had been achieved through millennia of adaptive evolution. While the late nineteenth century period of falling alluvial water tables, disastrous arroyo cutting, increased climatic amplitudes, severe drought, low temporal variability, high spatial variability, burgeoning regional population, and cultural diversity and stratification was an interval of resource stress, the responses to this crisis took place within the framework of a modern nation-state that was able to activate resources and processes that far surpassed the boundaries of the Southwest. Many of the technological and behavioral adaptational mechanisms that had evolved over the preceding three or four millennia were either submerged in the larger system or rendered irrelevant by permanently changed cultural-environmental circumstances.

CONCLUSIONS

The foregoing review of potential instances of resource stress caused by environmental, demographic, and behavioral variability during the last 2,000 years in the American Southwest provides a basis for assessing the relevance of Southwestern prehistory to understanding change and evolution in complex adaptive systems, in this case cultural systems. This objective is furthered by evaluating the Southwestern situation against Langton's seven principles of historical processes.

1. In the Southwest, the populations of interacting individuals and social units affected by these processes span levels of complexity ranging from the simplest households to subregional interaction systems such as the Chacoan system of the Colorado Plateau, the Hohokam and Salado systems of the Sonoran Desert and its borders, and the Casas Grandes system of the Chihuahuan Desert and adjacent areas. Adaptive culture change and evolution can profitably be studied at all these levels as long as the scale of complexity and the degree to which small-scale units are embedded in larger units are rigorously accounted for. Horizontal and vertical relationships and interactions within and among these units and interactions with external variables affect the evolutionary fitness of individual units at all levels of complexity and embeddedness. Ultimately, these relationships and transformations are responsible for the survival and evolution of societies and cultures.

   The maintenance and evolutionary transformation of social units of varying sizes and complexities occurs through the selection of attributes or traits associated with these units. Many of these attributes are behaviors or assemblages of linked behaviors involved in human individuals' and social units' relationships with the physical environment, one another, and other groups. Obviously, many other cultural phenomena are involved in structuring behavior, in linking individual behaviors into coherent systems and subsystems, and in integrating new behaviors into existing systems. Among these phenomena are organizational principles and relationships, belief systems, ideologies, iconographies, and many other tangible and intangible aspects of culture. Nonetheless, behavior is an important aspect, outcome, and expression of nonmaterial aspects of culture, is a primary focus of selective evolutionary pressures, and, as the preeminent creator of the archaeological record, is observable or inferable from the material remains and relationships of that record. Thus, the archaeological study of cultural evolution in the Southwest and elsewhere can logically and profitably be focused on relevant aspects of past human behavior that can be reconstructed from archaeological data.
2. The dictum that populations (social units) behave as units is true at all levels of complexity. In certain circumstances individuals behave as units relative to other individuals regardless of their culturally prescribed relationships. In other situations, these same individuals unite with fellow members of larger social units—such as households, lineages, clans, task groups, special-purpose

associations, and villages—in contraposition to similarly united groups of approximately equal scale. Hierarchically coordinated behavior of this sort continues up to the most complex social units such as multivillage communities, local amalgamations of such communities, regional interaction systems, tribes, confederations, and nations. Fortunately, this type of behavior is archaeologically discernible at most levels through the delineation of commonalities and boundaries within villages, communities, large-scale interaction networks, and regions. The identification of bounded social units allows the study of their cooperative, oppositional, and adaptive behavior.

3. Considerable disagreement exists whether higher levels of social complexity in the Southwest emerged spontaneously (that is, within the parameters of existing conditions, variations, and relationships) or were wholly or partly stimulated by outside influences (contacts with other, often "higher" cultures such as Mesoamerican states). The undeniable contribution of historical factors to particular instances of increased complexity does not exclude the possibility of self-generation in other cases of increased complexity or the possibility that degrees of self-generation were involved in cases characterized by outside stimuli. That social complexity arose rather rapidly on several occasions and that it involved the increasing specialization of individuals and social units is not in dispute. Increasing craft and role specialization of individuals and groups is manifest in the archaeological record as is the increasing size, differentiation, and hierarchy of residential, economic, socioreligious, and political groups (Upham 1982). Strong evidence for configurational similarities in many escalations of social complexity is an interesting aspect of Southwestern prehistory. The extent to which the common pattern of spatially discrete communities, each comprising a primary supraresidential site surrounded by numerous secondary settlements, organized into interaction networks by communications systems is an outcome of "universal" principles of cultural evolution, or is regionally specific, has important implications for the archaeological study of general cultural evolution.
4. Southwestern archaeology is unclear as to whether leaps in social complexity were characterized by initially increased diversity followed by increased uniformity. This indeterminance probably results from two factors: (i) the archaeological record has not been systematically examined for evidence of such changes in diversity, and (ii) the changes often may be so rapid as to leave few archaeological traces. In some instances of increased complexity, such as that of the thirteenth century in the Kayenta area, enhanced diversity (in both subsistence activities and settlement) may have preceded rather than accompanied attainment of the more complex Tsegi Phase configuration (Dean 1970, 1995a, 1995b). The strong Tsegi Phase pattern, in fact, may be a consequence of the reduction in variance associated with the selection of the most effective adaptive options. In any case, the Southwest, with its excellent preservation and

high-resolution temporal controls, should be an ideal venue for the archaeological investigation of the relationships between diversity and increased cultural complexity.

5. High-quality archaeological and chronological data sustain numerous demonstrations of the extensive circulation of innovations (discoveries, inventions, or borrowings), both material and nonmaterial, within and among prehistoric Southwestern societies. Thus, Southwestern archaeology illuminates the role of this principle in the evolution of cultural systems and should be ideally suited for elucidating the operation of this principle on both local and regional levels of analysis.
6. The approach to the study of sociocultural change and evolution embodied in the model of behavioral adaptation outlined above affirms that the evolution of Southwestern societies involved evolutionary processes operating on coupled fitness landscapes. The dynamic concept of carrying capacity integrates environmental, demographic, and behavioral variables in a manner that closely resembles the concept of coupled fitness landscapes in which changes in any aspect of the adaptive system alter relationships among all the components and reconfigure the landscape itself. The internal and external dynamics of these fluid situations provide fertile ground for the adaptive evolution of sociocultural systems.

The coupling of units on fitness landscapes to the extent that changes in the fitness of one unit alter the relative fitness of all other such units may provide a fruitful perspective on large-scale aspects of Southwestern prehistory. For example, this approach could provide a theoretical framework for understanding how the development of the Chacoan regional interaction system affected contemporaneous societies by altering the fitness landscape. Similarly, the idea that units of coupled fitness landscapes evolve at different and fluctuating rates, some changing while others remain "frozen" (Kauffman and Johnson 1992:344-346), could help explain the differential levels of societal complexity manifest across the region at any point in time. Finally, the concept of poised states may illuminate aspects of Southwestern prehistory. The ideas that ecosystems (including human behavioral components) exist in states ranging from chaos ("gas") to equilibrium ("solid"), that the transition phase ("liquid") from gas to solid is fraught with potential for evolutionary changes, and that minor perturbations at the liquid phase can cause evolutionary "avalanches" that propagate throughout the system (Kauffman and Johnson 1992:350–357), probably can help explain the simultaneous episodes of culture change that occurred at several times across the Southwest. Societies whose populations have breached carrying capacity limits are particularly vulnerable to even minor perturbations in any component of the adaptive system and may be said to exist in poised states. The widespread occurrence of behavioral changes throughout the region during such periods of resource stress may be cultural instances of evolutionary avalanches.

7. In the Southwest, as elsewhere, the operation of the historical principles drives sociocultural evolution, and the richness of the Southwestern archaeological and paleoenvironmental records allows the study of these processes in unprecedented detail.

In conclusion, human behavior in the southwestern United States during the last 2,000 years is amply suited for rigorous investigations of the details and processes of sociocultural evolution. Informed by theories relating the evolution of complex adaptive systems to a wide range of intrinsic and extrinsic variables and processes, investigators can use Southwestern archaeological data to evaluate hypotheses relating sociocultural change and evolution to various environmental, demographic, behavioral, and historical factors. An important aspect of this effort is the identification of critical points at which circumstances combined to produce situations with high potential for adaptive change. Chief among these are situations characterized by resource stress and/or economic uncertainty, both of which disrupt ecological and social relationships that had been achieved by previous adaptive adjustments. Within the parameters of the adaptational model, data on past environmental, demographic, and behavioral variability can be used to specify potentially critical points characterized by resource stress and economic uncertainty. Archaeological data may then be marshalled to investigate the responses to these stresses and the changes the responses effected in the cultures involved. Given the high quality of the archaeological data, the exceptional paleoenvironmental record, and the outstanding chronological controls, the Southwest is an ideal locus for investigating adaptive behavioral change over long time periods and using such studies to elucidate the evolution of complex adaptive systems.

ACKNOWLEDGMENTS

The perspective on behavioral adaptation outlined above was forged in intense discussions with numerous colleagues. The Gang of Five Plus One (Robert C. Euler, George J. Gumerman, Richard H. Hevly, Thor N. V. Karlstrom, and the late Fred Plog) generated numerous provocative ideas about paleoenvironmental reconstruction and its relevance to human behavior. School of American Research and Santa Fe Institute seminars on the "The Organization and Evolution of Prehistoric Southwestern Society" catalyzed many ideas on the relationships among environment, population, and culture. I am particularly indebted to the other members of the Environmental and Evolutionary Modeling Task Group at the first SFI Workshop (Patricia L. Crown, Stuart A. Kauffman, Christopher G. Langton, Roger Lewin, and W. H. Wills) and to Murray Gell-Mann for illuminating the application of current evolutionary theory to human culture as a complex adaptive system. The organizers of the School of American Research Advanced Seminar (Gumerman and Douglas

W. Schwartz) and Santa Fe Institute Workshop (Gell-Mann and Gumerman) provided inspiring and productive venues for exploring these topics. Joseph Tainter, organizer, and the other participants in the SFI Workshop on Resource Stress and Economic Uncertainty contributed much food for thought and pointed discussions of pertinent ideas. The late Donald A. Graybill and Gary S. Funkhouser helped create the Southwestern Paleoclimate Project (supported by Grants BNS-8818628 and DBS-9205968 from the Anthropology Program of the National Science Foundation) and produced numerous dendroclimatic reconstructions. Alison E. Dean drafted the illustrations, Figures 4, 8, and 9, from computer originals produced by Graybill and Funkhouser.

REFERENCES

Adams, E. Charles.

1991 *The Origin and Development of the Pueblo Katsina Cult.* The University of Arizona Press, Tucson.

Bak, Per, Chao Tang, and Kurt Wiesenfeld

1988 Self-Organized Criticality. *Physical Review A* 38:364–374.

Barnett, H. G.

1953 *Innovation: The Basis of Cultural Change.* McGraw-Hill Book Company, Inc., New York.

Boserup, Ester

1965 *The Conditions of Agricultural Growth: The Economics of Agrarian Change Under Population Pressure.* Aldine Press, Chicago.

Cordell, Linda S., and Fred Plog

1979 Escaping the Confines of Normative Thought: A Reevaluation of Puebloan Prehistory. *American Antiquity* 44:405–429.

Dean, Jeffrey S.

1970 Aspects of Tsegi Phase Social Organization: A Trial Reconstruction. In *Reconstructing Prehistoric Pueblo Societies*, edited by William A. Longacre, pp. 140–174. University of New Mexico Press, Albuquerque.

1984 Environmental Aspects of Modeling Human Activity-Locating Behavior. In *Theory and Model Building: Refining Survey Strategies for Locating Prehistoric Heritage Resources: Trial Formulations for Southwestern Forests*, edited by Linda S. Cordell and Dee F. Gree, pp. 8–20. Cultural Resources Management Report, No. 5. USDA Forest Service, Southwestern Region, Albuquerque.

1987 Southwest U. S.: The Archaic of Southern Arizona. Review of "The Cochise Cultural Sequence in Southeastern Arizona," by E. B. Sayles,

Anthropological Papers of The University of Arizona, No. 42 (1983), "Cultural and Environmental History of the Cienega Valley, Southeastern Arizona," by Frank W. Eddy and Maurice E. Cooley, *Anthropological Papers of The University of Arizona*, No. 43 (1983), and "The Geoarchaeology of Whitewater Draw, Arizona," by Michael R. Waters, *Anthropological Papers of The University of Arizona*, No. 45 (1986). *Quarterly Review of Archaeology* 8(4):10–14.

1988a A Model of Anasazi Behavioral Adaptation. In *The Anasazi in a Changing Environment*, edited by George J. Gumerman, pp. 25–44. Cambridge University Press, Cambridge.

1988b Dendrochronology and Paleoenvironmental Reconstruction on the Colorado Plateaus. In *The Anasazi in a Changing Environment*, edited by George J. Gumerman, pp. 119–167. Cambridge University Press, Cambridge.

1991 Social Complexity and Environmental Variability in Northeastern Arizona. Paper presented at the 24th Annual Chacmool Conference: "Culture and Environment: A Fragile Coexistence." The University of Calgary, 7 10 November 1991.

1995a Kayenta Anasazi Settlement Transformations in Northeastern Arizona: A.D. 1150–1350. In *Pueblo Cultures in Transition*, edited by Michael A. Adler. The University of Arizona Press, Tucson.

1995b Late Pueblo II-Pueblo III in Kayenta Branch Prehistory. In *Prehistoric Culture Change on the Colorado Plateau: Ten Thousand Years on Black Mesa*, edited by Shirley L. Powell and F. E. Smiley. The University of Arizona Press, Tucson.

Dean, Jeffrey S., William H. Doelle, and Janet D. Orcutt

1994 Adaptive Stress, Environment, and Demography. In *Themes in Southwest Prehistory*, edited by George J. Gumerman, pp. 53–86. School of American Research Press, Santa Fe.

Dean, Jeffrey S., Robert C. Euler, George J. Gumerman, Fred Plog, Richard H. Hevly, and Thor N. V. Karlstrom

1985 Human Behavior, Demography, and Paleoenvironment on the Colorado Plateaus. *American Antiquity* 50:537–554.

Dean, Jeffrey. S., and Alexander J. Lindsay, Jr.

1978 Special Use Sites in Long House Valley, Northeastern Arizona: An Analysis of the Southwestern Anthropological Research Group Data File. In *Limited Activity and Occupation Sites: A Collection of Conference Papers*, edited by Albert E. Ward, pp. 109–117. Center for Anthropological Studies Contributions to Anthropological Studies, No. 1. Albuquerque.

Dean, Jeffrey S., Alexander J. Lindsay, Jr., and William J. Robinson

1975 The Long House Valley Project. Paper presented at the 40th Annual Meeting of the Society for American Archaeology, Dallas.

1978 Prehistoric Settlement in Long House Valley, Northeastern Arizona. In *Investigations of the Southwestern Anthropological Research Group: An*

Experiment in Archaeological Cooperation: The Proceedings of the 1976 Conference, edited by Robert C. Euler and George J. Gumerman, pp. 25–44. Museum of Northern Arizona, Flagstaff.

Dean, Jeffrey S., and William J. Robinson

1977 *Dendroclimatic Variability in the American Southwest,* A.D. *680–1970.* U.S. Department of Commerce, National Technical Information Service, PB-266 340. Springfield, Virginia.

1978 *Expanded Tree-Ring Chronologies for the Southwestern United States.* Laboratory of Tree-Ring Research, The University of Arizona, Tucson.

Dunnell, Robert C.

1980 Evolutionary Theory in Archaeology. In *Advances in Archaeological Method and Theory*, Vol. 3, edited by Michael B. Schiffer, pp. 35–99. Academic Press, New York.

Eddy, Frank W., and Maurice E. Cooley

1983 Cultural and Environmental History of Cienega Valley Southeastern Arizona. In *Anthropological Papers of The University of Arizona*, No. 43. The University of Arizona Press, Tucson.

Graybill, Donald A.

1989 The Reconstruction of Prehistoric Salt River Streamflow. In *The 1982-1984 Excavations at Las Colinas: Environment and Subsistence*, by Donald A. Graybill, David A. Gregory, Fred L. Nials, Suzanne K. Fish, Robert E. Gasser, Charles H. Micsicek, and Christine R. Szuter, pp. 25–38. Arizona State Museum Archaeological Series, No. 162, Vol. 5. Tucson.

Gumerman, George J. (editor)

1988 *The Anasazi in a Changing Environment.* Cambridge University Press, Cambridge.

1994 *Themes in Southwest Prehistory.* School of American Research Press, Santa Fe.

Gumerman, George J., and Murray Gell-Mann (editors)

1994 *Understanding Complexity in the Prehistoric Southwest.* Santa Fe Institute Studies in the Sciences of Complexity, Proc. Vol. XVI. Addison-Wesley Publishing Company, Reading.

Hassan, Fekri A.

1978 Demographic Archaeology. In *Advances in Archaeology Method and Theory*, Vol. 1, edited by Michael B. Schiffer, pp. 49–103. Academic Press, New York.

Hevly, Richard H.

1988 Prehistoric Vegetation and Paleoclimates on the Colorado Plateaus. In *The Anasazi in a Changing Environment*, edited by George J. Gumerman, pp. 92–118. Cambridge University Press, Cambridge.

Karlstrom, Thor N. V.

1988 Alluvial Chronology and Hydrologic Change of Black Mesa and Nearby Regions. In *The Anasazi in a Changing Environment*, edited by George J. Gumerman, pp. 45–91. Cambridge University Press, Cambridge.

Kauffman, Stuart A.
1991 Antichaos and Adaptation. *Scientific American* 265(2):78–84.
Kauffman, Stuart A., and Sonke Johnson
1992 Co-Evolution to the Edge of Chaos: Coupled Fitness Landscapes, Poised States, and Co-Evolutionary Avalanches. In *Artificial Life II: Proceedings of the Workshop on Artificial Life Held February, 1990, in Santa Fe, New Mexico*, edited by Christopher G. Langton, Charles Taylor, J. Doyne Farmer, and Steen Rasmussen, pp. 325–369. Santa Fe Institute Studies in the Sciences of Complexity, Proceedings Volume X. Addison-Wesley Publishing Company, Reading.
Kauffman, Stuart, and Simon Levin
1987 Towards a General Theory of Adaptive Walks on Rugged Landscapes. *Journal of Theoretical Biology* 128:11–45.
Kauffman, Stuart A., and Edward D. Weinberger
1989 The NK Model of Rugged Fitness Landscapes and Its Application to Maturation of the Immune Response. *Journal of Theoretical Biology* 141:211–245.
Kirch, Patrick V.
1980 The Archaeological Study of Adaptations: Theoretical and Methodological Issues. In *Advances in Archaeological Method and Theory*, Vol. 3, edited by Michael B. Schiffer, pp. 101–156. Academic Press, New York.
Leonard, Robert D.
1989 Resource Specialization, Population Growth and Agricultural Production in the American Southwest. *American Antiquity* 54:491–503.
Leonard, Robert D., and George T. Jones
1987 Elements of an Inclusive Evolutionary Model for Archaeology. *Journal of Anthropological Archaeology* 6:199–219.
Matson. R. G.
1991 *The Origins of Southwestern Agriculture.* The University of Arizona Press, Tucson.
Plog, Fred, George J. Gumerman, Robert C. Euler, Jeffrey S. Dean, Richard H. Hevly, and Thor N. V. Karlstrom
1988 Anasazi Adaptive Strategies: The Model, Predictions, and Results. In *The Anasazi in a Changing Environment*, edited by George J. Gumerman, pp. 230–276. Cambridge University Press, Cambridge.
Plog, Stephen
1986 Patterns of Demographic Growth and Decline. In *Spatial Organization and Exchange: Archaeological Survey on Northern Black Mesa*, edited by Stephen Plog, pp. 224–255. Southern Illinois University Press, Carbondale and Edwardsville.

Powell, Shirley
1988 Anasazi Demographic Patterns and Organizational Responses: Assumptions and Interpretive Difficulties. In *The Anasazi in a Changing Environment*, edited by George J. Gumerman, pp. 168–191. Cambridge University Press, Cambridge.

Sayles, E. B.
1983 The Cochise Cultural Sequence in Southeastern Arizona. In *Anthropological Papers of The University of Arizona*, No. 42. The University of Arizona Press, Tucson.

Schoenwetter, James
1962 The Pollen Analysis of Eighteen Archaeological Sites in Arizona and New Mexico. In *Chapter in the Prehistory of Eastern Arizona, I*, by Paul S. Martin, John B. Rinaldo, William A. Longacre, Constance Cronin, Leslie G. Freeman, Jr., and James Schoenwetter, pp. 168–209. Fieldiana: Anthropology, Vol. 53. Chicago Natural History Museum, Chicago.

Smiley, Frances Edward IV
1985 *The Chronometrics of Early Agricultural Sites in Northeastern Arizona: Approaches to the Interpretation of Radiocarbon Dates.* Ph.D dissertation, University of Michigan, Ann Arbor. University Microfilms, Ann Arbor.

Upham, Steadman
1982 *Polities and Power.* Academic Press, New York.

Vivian, R. Gwinn
1990 *The Chacoan Prehistory of the San Juan Basin.* Academic Press, Inc., San Diego.

Waters, Michael R.
1986 The Geoarchaeology of Whitewater Draw, Arizona. In *Anthropological Papers of The University of Arizona*, No. 45. The University of Arizona Press, Tucson.

Wills, W. H.
1988 *Early Prehistoric Agriculture in the American Southwest.* School of American Research Press, Santa Fe.

Wills, W. H., Patricia L. Crown, Christopher G. Langton, and Jeffrey S. Dean
1994 Complex Adaptive Systems and Southwestern Prehistory. In *Understanding Complexity in the Prehistoric Southwest*, edited by George J. Gumerman and Murray Gell-Mann, pp. 297–339. Santa Fe Institute Studies in the Sciences of Complexity, Proc. Vol. XVI. Addison-Wesley Publishing Company, Reading, MA.

Paul E. Minnis
Department of Anthropology, University of Oklahoma, Norman, OK 73019

Notes on Economic Uncertainty and Human Behavior in the Prehistoric North American Southwest

INTRODUCTION

All human populations have faced problems of economic uncertainty, and all have developed a range of ingenious behaviors to reduce risk. Here I focus upon food provisioning problems as an important nexus of uncertainty and risk in the prehistoric North American Southwest. The archaeological study of such responses has value for explaining prehistory, among other uses. We can build more accurate models and theories of human behavior under food shortages, especially long-term cultural changes best observed through the archaeological record.

Despite a number of reasonably sophisticated studies exploring the relationships between environmental variability and prehistoric human behavior in the North American Southwest (e.g., Dean et al. 1985; Euler et al. 1979; Gerald 1976; Gumerman 1988; Minnis 1985a; Rautman 1990; Reid 1978; Spielmann 1982), there seems a widespread fatigue with the analysis of prehistoric environmental variation in human prehistory. Reid (1978:195) noted that, "there is a nagging uneasiness these days that archaeologists everywhere, and especially in the North American Southwest, have been too quick to clutch at environmental variability as the causal agent responsible for variability in culture and behavior." All too often a period of unusual environmental variability is used by itself to explain a concurrent cultural change. A simple correlation between environment and culture, however, is an insufficient explanation. On the other hand, southwestern archaeologists would be foolish to assume that environmental fluctuations were irrelevant to prehistoric communities. In order to develop better models of the relationships between prehistoric behavior and economic uncertainty, we need to pay closer attention to the types of cultural responses to environmental change documented in the ethnographic record. What do people do when faced with significant environmental fluctuation? Furthermore and more importantly, there may be an ordered sequence of responses that, if documented in the archaeological record, increases confidence in our interpretation of the past.

With this in mind, I will first review various responses to food acquisition problems. Following this will be a discussion of how responses are related to each other, and I will propose that the use of these strategies is patterned in a sequential order. Then I discuss the relationships between food scarcity and sociocultural change, providing what may be a counterintuitive view that all responses, not just the most severe responses, can contribute to social and cultural change.

Archaeological study of prehistoric economic strategies in general and responses to resource stress in particular is useful to more than archaeologists; it can also have a more practical value. What appear to southwestern archaeologists as simple descriptions of risk-reduction strategies can be enormously valuable to those working to understand and implement food relief efforts. People involved in largely subsistence economies throughout the world are experiencing serious food provisioning problems, and there is little reason to expect that these problems will abate; in fact, they are likely to become even worse. Sustainable agriculture in many areas remains an elusive goal. Not only are populations increasing in many areas of the world, but economic relationships and natural environments are deteriorating. Novel productive strategies are needed, and, fortunately, the past can be a source of innovation. A multitude of risk-avoidance strategies developed by human populations is discernible only or largely through the archaeological record, including in the North American Southwest. The cultural and environmental diversity of the prehistoric North American Southwest was greater than the region's ethnographic and historic diversity, and therefore should offer more examples of responses to food provisioning problems. Novel responses, such as unique drought-resistant agricultural technology and crops, documented archaeologically could well provide models for modern development and disaster relief efforts.

One example will suffice to illustrate this point. At least one species of cultivated century plant (*Agave*) with the associated technology of production and processing has been noted at several locations in central Arizona (Fish et al. 1985). This discovery documents a previously unknown agricultural complex that has stimulated research by agronomists. Many arid regions, including the North American Southwest, are having increasing problems of water availability for agriculture due to decreased groundwater reserves, decreased water quality, more water competition between rural and urban populations, and overallocation of surface water.

Under such circumstances, knowledge of how prehistoric farmers coped with low precipitation may become increasingly useful.

Low precipitation was a major factor in prehistoric human ecology in the North American Southwest. There are four reasons to focus on food shortages caused by low precipitation. First, ethnohistorical documentation shows that low precipitation has been the major cause of low crop yields in parts of the North American Southwest (Stephens 1936:1940). Records of historic dry farming have reported poor to disastrous yield as frequently as 20 to 30 percent of the years (e.g., Stanton et al. 1932). Low moisture is the primary limiting factor of agricultural production in many other semiarid regions, such as Mexico (Kirkby 1973, 1978), Greece (Forbes 1989), India (e.g., Marten and Saltman 1986), and Africa (Akong'a 1988; Horowitz 1976; McLoughlin 1970). Second, the importance of drought should not surprise us given that much of the inhabitable North American Southwest has an arid to semi-arid climate. As a general rule, precipitation variation is greater under low moisture regimes than in wetter climates (e.g., Tuan et al. 1973). Third, low precipitation also affects the abundance and availability of noncultivated foods, an especially important source of sustenance during food shortages. Biological productivity, measured for example by net primary productivity, is strongly and nearly linearly correlated with precipitation in semiarid to arid environments (e.g., Noy-Meir 1985). Fourth, we are fortunate to have the best prehistoric precipitation and paleoenvironmental data anywhere in the world. We also possess some of the most reliable and precise archaeological data available.

The fact that I emphasize low precipitation does not mean that other factors, both environmental and social/cultural, do not cause low food availability. Food scarcity can easily be the consequence of a variety of factors such as conflict. Nor am I arguing that food shortages are not interdependent and interactive with other variables. For example, hunger seasons in arid to semiarid temperate regions, such as the North American Southwest, tend to be rather benign for human cultures and populations, although at times not so for individuals. In humid tropical regions, in contrast, hunger seasons tend to co-occur with the period of heaviest disease load. This difference alters relationships among the availability and allocation of labor, subsistence, and dietary requirements.

RESPONSE TYPES

There are numerous individual responses to food scarcity (also called buffering mechanisms, coping strategies, etc.). Unfortunately, most ethnographic and historical examples of community responses to food shortages are anecdotal; that is, the ecological and cultural contexts within which these actions occur are rarely presented. Nor are descriptions of the relationships among different strategies documented. Therefore, I will discuss individual response types. Here I will largely ignore physiological actions and concentrate on human behavior.

Numerous behavioral responses have been described, and a variety of classifications of responses has been proposed. Colson (1980:21), for example, enumerated five of what she called "devices":

> (1) diversification of activities rather than specialization or reliance on a few plants or animals, (2) storage of foodstuffs, (3) storage and transmission of information on what we can call famine foods, (4) conversion of surplus into durable valuables which could be stored and traded for food in an emergency, and (5) cultivation of social relationships to allow one to tap resources of other regions.

More recently, Halstead and O'Shea classified coping strategies into four more inclusive categories. These are: mobility, diversification, physical storage, and exchange that "exploit[s] favourable aspects of temporal and spatial structure of variability to mitigate the risk of scarcity" (1989:34). These categories clearly overlap with Colson's, with the exception of mobility, which Colson did not include. I will summarize categories using Halstead and O'Shea's classification while discussing some specific responses that they do not consider.

MOBILITY

Mobility is a set of strategies that increases the spatial resource base. The geographical scale of mobility is determined as much by the social landscape as by the natural environment. Clearly the effectiveness of mobility is to a degree dependent on population density. Under low population densities relative to the regional resource structure, this strategy is quite effective. In fact it should be the response of choice. With increasing population densities, especially accompanied with strengthened rules of land tenure, access to distant terrestrial (as opposed to marine) resources may require establishment of economic/social relationships with distant groups, or alternatively simple brute force, to maintain access to these resources.

The relationships among population density, the resource structure, and mobility can be quite complex and are not simply a density-dependent process. Although tempting to posit, movement under economic uncertainty is not always from high-population-density areas to low-population-density areas. The Enga, a group of communities totaling 150,000 people in highland Papua New Guinea, provide a telling example. The Fringe Enga (Waddell 1972, 1975) are upland farmers whose agriculture is extremely susceptible to the effects of frost, and they have developed several ingenious farming tactics to minimize frost damage. The related Valley Enga live in more densely packed lowland valleys, and their economic production is more stable. Waddell noted that one of the major responses by the Fringe Enga to a devastating frost in 1972 (as well as during two earlier periods of agricultural failure remembered by informants) was for a significant portion of the population to migrate to Valley Enga communities for at least several years, and for some, indefinitely. Phrased in the crudest terms, population movement was from low- to high-density areas. This demographic movement into already crowded regions was not passive but rather was encouraged; "... in effect, individual groups among the Central Enga actively solicit immigrants in spite of the fact that overall densities and pressures on resources are high compared with the fringe areas" (Waddell 1975:267). People were needed to maintain Valley Enga populations in the face of intergroup conflict and other social processes.

One issue related to mobility as a response to food acquisition problems is the effects of human exploitation on outlying regions. What are the ecological effects of intermediate to dense populations on large underexploited and weakly claimed territory (often called buffer zones). Does the much touted "tragedy of the commons" (Hardin 1968; Hardin and Badens 1977; McCay and Acheson 1987) occur under these circumstances with the resulting environmental deterioration of the outlying resource zones? Was Aristotle correct when he stated, "that which is common to the greatest number has the least care bestowed upon it"? (cited in McCay and Acheson 1987:2). Is the supposed environmental neglect of the underclaimed or commonly claimed territory intensified during food scarcity through such processes as overhunting or overharvesting? This topic, and the more general topic of anthropogenic environmental change, have been largely ignored by southwestern archaeologists.

TYPES OF DIVERSIFICATION

Diversification includes a wide range of strategies. The logic of diversity as a risk-reduction strategy is quite clear to us; our folk wisdom admonishes us "not to put all our eggs in one basket." Here I will break diversification down into three types: (1) polyculture, (2) multiple field locations/production technology, and (3) low-preference foods.

POLYCULTURE. One of the most common strategies for reducing risk is growing a large range of crops and cultivars. This approach is usually linked to the second type of diversification, multiple farming techniques and multiple field locations. Polyculture was practiced in the prehistoric and historic North American Southwest. The large number of significant Mesoamerican crops that were ultimately grown in the prehistoric North American Southwest and the rapid addition of many Eurasian crops after European contact attest to the receptivity of southwestern farmers to the incorporation of new cultigens and broadening their crop base (Ford 1981).

There may well have been regional differences in crop diversity within the prehistoric North American Southwest. Crop assemblages on the Colorado Plateau and perhaps in the Mogollon areas appear to have been less diverse than elsewhere in the prehistoric North American Southwest (Ford 1981). Some Sonoran Desert and Transitional Zone (the mountainous zone between the Sonoran desert and the Colorado Plateau/Mogollon Highlands) populations, and perhaps groups in northern Mexico, seem to have had a greater range of crops. It is quite conceivable that low crop diversity on the Colorado Plateau increased vulnerability to low precipitation, compared to some other archaeological regions in the North American Southwest (as well as the Midwest and Southeast of North America, and even Mesoamerica).

MULTIPLE FIELD LOCATIONS/PRODUCTION TECHNOLOGY. It is very common for agriculturalists, including those in the North American Southwest, to cultivate many fields, often in different microenvironmental settings and using different agricultural technology, each of which might be susceptible to different risk factors or which might have different probabilities of risk. Among other things this spreads spatial risk: a factor affecting the yield of crops in one location may not affect crops in other locations. Hopi agriculture is a reasonably well-documented example from the North American Southwest (Bradfield 1971; Whiting 1939).

One problem with multiple field locations is increased logistical costs. I do not know what a calculation of the logistical costs of multiple field locations would have been in the prehistoric North American Southwest, although one could recalculate Lightfoot's (1979) figures of energetic efficiency of food transportation to get an idea of logistical costs. McCloskey (1976) estimated that the cost of multiple field locations in the English common field agriculture system prior to 1700 was on the order of a 10% drop in production efficiency. Furthermore, field scattering can require land tenure over large areas, with possible increase in conflict over territoriality with other groups.

LOW-PREFERENCE FOOD/ALTERNATIVE FOODS/FAMINE FOODS. Perhaps one of the most common, least understood, and least appreciated coping strategies is the use of alternative foods, usually low-preference foods that would not otherwise be consumed. One case illustrated what I believe to be the largest number of famine foods used, even though such foods are often ignored in the ethnobiological literature. The "Chiu-Huang Pen-ts'ao," published in 1559, lists 414 plant famine foods recorded for Hunan, China, alone (Read 1946). The study of famine food use in the North American Southwest provides clues to changing subsistence activities (Minnis 1991).

The use of famine foods is a complex phenomenon involving a multitude of both biological and cultural factors. Famine food can include agricultural by-products not normally consumed, seed stock, and resources used only when more preferred foods are unavailable, either as seasonal hunger plants or only during famines. The status of a resource as a famine food is not based solely on its biology and biochemical profile, but also involves a range of social, cultural, political, and economic factors.

In light of the hundreds of useful plants recorded in the ethnobotanical literature of the Desert Borderlands, relatively few famine foods are known. I document about forty (see Minnis 1991:Table 1). Undoubtedly there are many unreported resources which have been used as famine foods by the indigenous populations in the region, and it is quite likely that much of the knowledge of traditional famine foods has been lost since European contact and before intensive ethnographic documentation began around one hundred years ago.

An additional dimension may also be an important reason for the few famine foods recorded. By the time ethnographers recorded the ethnobotany of most native peoples in the area, many of these peoples had dramatically altered their subsistence base. Such changes included the introduction of new crops, new agricultural technology, animal husbandry, increased sedentism, and substantial involvement in market economies. I suggest that traditionally important foods had been relegated to less frequent use. This conclusion is based on a comparison of prehistoric foods common in coprolites from the northern North American Southwest with ethnographically documented famine foods, especially those recorded for the Hopi (Minnis 1991). Some of these traditional wild food plants then became potential famine foods. In short, the introduction of a new set of resources caused a resorting of general food preference patterns, with some newly acquired plants replacing some former foods and the latter in turn becoming less commonly used. Previous famine foods were thus replaced by what were once more common foods. The knowledge of the "original" famine foods may have been completely lost. Specifically, many of the famine foods recorded for modern groups in the Desert Borderlands may well have been seasonal hunger foods in the past. Robbins, Harrington, and Freire-Marreco, for example, obliquely narrate this shift for the Tewa of San Juan Pueblo,

> But nowadays, although wild berries and nuts are still gathered in autumn and green weeds are eagerly sought and eaten in the spring, there is a very general and increasing neglect of all but *the most common and best-liked.* Formerly it was a matter of necessity that the housewife should know them and store them; for although in normal years they were merely a pleasant addition to the diet, yet drought, flood, fire, or a hostile raid might destroy the crops at any time, thus making the wild products an indispensable resource (Robbins, Harrington, and Freire-Marreco 1916:76; emphasis added).

How is this infrequently used and specialized knowledge of famine foods maintained within the pool of knowledge shared by a group? The simple observation of behavior is effective for learning about common foods, including foods used during yearly hunger seasons, but it could well be ineffective for infrequently used foods, if severe shortages are less frequent than once per generation. If so, then other mechanisms of learning may be particularly important for famine food use. As many have pointed out, myths, legends, rituals, and stories about previous food shortages are critical for transmitting knowledge of famine food use (e.g., Colson 1980; Cove 1978; Galt and Galt 1979; Marcus 1982; Reining 1970; Roys 1967). Thus, oral tradition may be especially important in perpetuating knowledge of famine foods. Special attention should focus on women's knowledge, because they seem to have the greatest familiarity with famine foods (Ali 1984). Yet, the role of male secular and ritual knowledge of plant foods cannot be ignored, as will be seen in the following Zuni example.

A Zuni example documents how ritual knowledge can perpetuate information for successful coping strategies. Bunzel has provided a translation of *Sayataca's night chant*. In this prayer, a range of edible plants is enumerated after mention of cultigens:

> ...and then the seeds of the piñon tree, the seeds of the juniper tree, the seeds of the oak tree, the seeds of the peach tree, the seeds of the black wood shrub, the seeds of the first flowering shrub, the seeds of the *kapuli* shrub, the seeds of the large yucca, the seeds of the branched yucca, the seeds of the brown cactus, the seeds of the small cactus, and then also the seeds of the wild grasses—the evil smelling weeds, the little grass, *tecukta*, *kucutsi*, *o'co*, *apitalu*, *sutoka*, *mololoka*, *piculiya*, small *piculiya*, *hamato*, *mitaliko*, and then also the seeds of those that stand in their doorways, namely the cat-tail, the tall flags, the water weeds, the water cress, the round-leafed weed... (Bunzel 1932:714).

R. Ford (1988) points out that seeds of these native edible plants are a part of some Zuni ritual paraphernalia. These plants must be collected each year, and this maintains the knowledge of the use of native plants and collection locations of various foodstuffs which might otherwise be ignored.

Clearly, use of alternative foods was an important response to food shortages by prehistoric peoples in the North American Southwest, yet it is more difficult to study than one might first believe. One archaeological implication of this study (and of course assuming that my arguments are correct) is the need for judicious use of ethnobotanical data available in the ethnographic record. While similar plants are represented in the archaeological and ethnohistorical periods, their structural role in subsistence may well have been different.

PHYSICAL STORAGE/SURPLUS PRODUCTION

Storage is one of the most widespread and effective strategies for reducing the effects of short-term food scarcity. It requires the production of surpluses that reduce temporal variation and to a lesser degree spatial variation in food production. There is a great deal of research on prehistoric storage, and other chapters in this volume will deal with storage in greater detail. However, one point should be emphasized. Storage need not be limited to commodities directly consumed. As Colson (1980) and many others have discussed, surplus can be converted into durable goods (and livestock in the Old World) that can be reconverted ("sold") into foodstuffs, although the exchange value can be reduced significantly.

EXCHANGE/SOCIAL INTERACTION

The importance of social networks as a coping strategy for human populations has been recognized through a number of theoretical approaches (e.g., economic anthropology [Colson 1980]; ecological anthropology [Rappoport 1979]; sociobiology/behavioral ecology [Cashdan 1990]; and archaeology [Braun and Plog 1982]). The basic argument for the role of social relations as a coping strategy can be summarized briefly: "In the absence of unrestricted mobility, social groups faced with food provisioning problems will have to enlarge their social/economic network so as to have access to a more reliable food supply" (Minnis 1985a:20). Or to rephrase it in Halstead and O'Shea's terms (although they might not agree with the rephrasing), one might say that ultimately the exploitation of spatial variability is more reliable than exploitation of temporal variability. I would argue, as have others, that while social and economic relationships may be the most effective strategy for risk reduction, they also carry the greatest cost in the form of obligations to others. Individuals, families, or kin groups participating in reciprocal relationships can no longer be concerned solely with their own interests and survival without risking violation of social norms. On this point, the views of sociobiologists and evolutionary and economic anthropologists tend to converge. Since many other papers in this volume focus on exchange as a coping strategy, I will not consider this further.[1]

One other type of social interaction must be considered. Warfare and raiding could be considered a form of either diversification or social interaction. This response type is often ignored by southwestern archaeologists. Perhaps because of its political implications, negative reciprocity tends to be ignored as a social process in the prehistoric North American Southwest. Additionally, low intensity conflict can be difficult to see in the archaeological record. I am as guilty in this regard as anyone else. Part of this neglect may be due to the basic ineffectiveness of raiding in nonhierarchical societies. Raiding parties may have difficulty carrying back large amounts of food, and raiding invites retaliation.

The ubiquity and intensity of conflict in the prehistoric North American Southwest is uncertain, especially among the Anasazi (Haas 1989). Interesting arguments have been made, however, for widespread conflict in some areas during late prehistory in the southern North American Southwest (Di Peso 1974; Fish and Fish 1989; Wilcox 1989). The utility of warfare and raiding as a response to food shortages may have increased with greater sociopolitical complexity. Perhaps we could even conceive of the development of regional hegemony with tribute demands (tribute being a form of institutionalize "raiding") by complex regional polities in the North American Southwest.

[1] See the chapters by Kohler and Van West, Tatman, and Hegmon.

OTHER COPING STRATEGIES

There are numerous other responses. Perhaps the simplest strategy, and the one with the most limited effectiveness, is resource conservation. Consumers can often reduce their food consumption. There are, however, well-known biological limits to dietary reduction, after which health and the availability of labor are diminished. Conservation is best known as a strategy for coping with expected and short hunger seasons.

Changes in ritual activity are recorded to occur during food shortages. Such alterations can include intensification of ceremonies, change in ritual types, or attenuation of ritual content (e.g., Cove 1978).

The effects of food deprivation can been focused on specific age cohorts. Turnbull (1972, 1978) graphically described how reproducing-age Ik remained relatively well fed compared with the old and young.

There are several ways for social groups to remove members, thus reducing the food needed to support the group. Dependents can be married off, sold, or simply "turned loose." Forbes (1989) described early and otherwise undesirable marriages as a way of reducing family size among Greek peasants.

Few ethnographers consider economic specialization as a response to food shortages, because it seems to contradict the general principle of diversification. There are examples of communities poorly situated for adequate food production that produce crafts for exchange with other villages. Picuris Pueblo in New Mexico (Ford 1972) and Chamula in southern Mexico (Collier 1975) are two examples. Increased craft production is recorded to have occurred during food shortages (e.g., the Gwembe Tonga of Zimbabwe), although one wonders whether this behavior is more frequent in communities with extensive markets.

A response commonly recorded among current populations is migration to cities or other locales to obtain wage labor, although bottlenecks in agricultural labor needs occur. It is unknown how common this strategy was prehistorically, and it is unlikely to have occurred widely in the prehistoric North American Southwest.

Finally, many models treat intensification of economic activities as a result of increasing imbalances between population size and production. There are ethnographic examples of economic intensification during food shortages (e.g., Maclachlan 1983). This process, however, is most widely discussed in regard to the first use of agriculture (Berry 1982, Ford 1985; Matson 1992; Minnis 1985b; Wills 1988).

PATTERNS OF RESPONSES

Documenting the presence of food shortages or stress in the archaeological record is difficult. This is particularly true if cultural systems respond to food stress in the absence of documented biological markers of nutritional problems. That is, cultural behavior can be more sensitive to food deprivation than physiology. The fact that risk-reduction responses known ethnographically to have been used during food shortages are present in the archaeological record is helpful but not conclusive, because many individual coping strategies are effective for a range of problems. For example, families or communities can migrate, have conflict, or intensify their economic activities for many reasons. The analysis of economic uncertainty in prehistory can be strengthened if the patterning of these coping behaviors can be demonstrated in the archaeological record. Therefore, models of response patterning should be of special interest to archaeologists.

Numerous scholars using different perspectives have proposed a sequential patterning of responses to food shortages; strategies to buffer food provisioning problems are not randomly used. In their own ways, each of these approaches suggests that the magnitude of the response should match the severity of the problem. Five are briefly outlined here. The first is rooted in the work of the evolutionary ecologist Slobodkin (1964, 1968; Slobodkin and Rapoport 1974), which I used as the base for my study of food stress among the prehistoric occupants of the Rio Mimbres region of the North American Southwest (Minnis 1981, 1985). Second, Halstead and O'Shea (1989), likewise, modeled the relationship between risk and social change and outlined a scheme of responses. Third, Forbes (1989), a cultural anthropologist working with modern Greek peasants, proposed a related model of human responses to economic uncertainty. Waddell (1975), a cultural geographer, presented a fourth example of a hierarchical model of response sequence. Fifth and finally, Watt (1988) outlines a sequential model for Hausa responses to drought. Although not discussed here, Rudel (1980) also suggested a sequential response order to resource scarcity, in his case the gasoline shortage of 1973–1974 in Atlanta, Georgia. It may well be that the sequential ordering of responses to perturbations is a general characteristic of all complex adaptive systems.

Slobodkin and Rapoport summarize the basis of their model as follows:

> Successful evolution requires the maintenance of flexibility in the response to environmental perturbations and that this flexibility must be maintained in the most parsimonious way. The parsimony argument is that organisms must not make an excessive or unnecessary commitment in responding to perturbations, but at the same time the deeper responses must be ready to take over to the degree that superficial responses are ineffective (Slobodkin and Rapoport 1974:198).

It is necessary for this view to have definitions of "depth" and "superficiality." Slobodkin and Rapoport proposed three criteria to "measure" depth: the speed of activation, the amount of resource commitment, and the reversibility of the response. Six levels of response were ranked according to increasing response depth: behavior, physiology, physiological acclimatization, death rate changes, selective mortality and fecundity, and genetic changes.

For a variety of reasons discussed elsewhere in greater detail (Minnis 1985a), I used Slobodkin's basic perspective and modified it in a way that I felt better suited human populations. I argue that social relations are the most effective response for the most serious problems. Yet the cost is high to families, because more socially inclusive responses tender obligations to more groups, thus reducing the ability of families to focus on their own well being. Additionally, more socially inclusive responses should be less reversible, because they involve greater numbers of social groups. Specifically, I proposed that risk-reduction responses involving multiple social groups with be ordered so that more socially inclusive responses will be used only after less inclusive responses. This model was then compared with (1) three ethnographic descriptions of response sequences (the Mae Enga of Papua New Guinea, the Gwembe Tonga of Zimbabwe, and the Tikopians of the Pacific) and (2) the archaeological sequence of the Rio Mimbres Valley in southwestern New Mexico. In a general sense, the model of increasing social inclusivity is consistent with these examples, although the archaeological case study was based on less than optimal data.

Halstead and O'Shea's (1989:4) perspective mirrors mine closely. They propose that "societies deploy an array of different strategies in a hierarchy of responses, which are equated with both the scale of the producing and consuming units (individuals, households, villages, states) and the magnitude of the resource failure encountered." They further suggest that there should be strong selective pressure to embed high-level responses within the cultural system. These behaviors then become largely irreversible, and in the event of a resource failure, they can cause severe social disruption. As an example, high-level behaviors provide surplus production for ceremonial events. Although not a high-level response, the example of Zuni famine food knowledge perpetuated within ritual contexts is an obvious example of embeddedness.

Forbes (1989) divided what he terms Hazard Response Mechanisms into three categories. First-defense mechanisms include polyculture, fragmentation, and overproduction. Storage is the example of his second category, safety-net mechanisms. The third category, emergency mechanisms, includes begging, use of low-preference foods (such as animal fodder), and marrying off eligible dependents.

Forbes (1989:95) proposed differences between lower-level and higher-level responses. He explicitly analyzed only household-level responses. Even though the population he studied is part of a nation-state, he argued that the Greek state is less involved in risk reduction for this rural community than many would believe. He outlined the following characteristics of lower-level responses: (1) continuous or frequent operation, (2) energetically expensive (security instead of productivity), (3) well integrated into other aspects of culture, and (4) therefore low visibility to both participant and observer. In contrast, he suggests that higher-level responses share the following characteristics: (1) infrequent use, (2) low energy expense, (3) high social expense, since they are counter to social rules, and (4) therefore high visibility to participant and observer alike. The key characteristics to define the response categories, according to Forbes, much like the other models discussed, are frequency of operation, severity/economic cost, and social "cost."

Waddell (1972, 1975) studied the Enga of New Guinea. He outlined a similar, although more attenuated, model. Not unexpected for a geographer, responses are partitioned on the basis of space. He posits three levels of responses—local, interregional, and extraregional—which he suggests are linked in such a way that, "the lowest (local) level is in constant operation, whereas the other two become progressively operational as the intensity of the climatic perturbation (frost) increases" (Waddell 1975:267).

Watt (1988) analyzed the responses to drought in the 1970s by the Hausa of Nigeria, whose farming success is largely dependent on rainfall. He concludes that the structure of Hausa responses maintain "adaptive flexibility." Consistent with the other examples cited here, Watt further argues that responses, from the use of famine foods on one extreme to permanent out-migration on the other, are order according to two criteria: increasing commitment of domestic resources and decreasing reversibility. Yet, "the precise constellation of responses during a crisis, and in particular the decision to migrate, reflects in large part local income-earning opportunities" (Watt 1988:273). Like Greek peasants, the Hausa are more fully embedded within a market economy than the other examples.

The key element for all of these views is that responses are not randomly used. Explicitly or implicitly, each model argues for a sequence of responses with "resource cost" and "perturbation severity" (however defined and however measured) inversely related. While there is a theoretical unity in the general views outlined above, the approaches do diverge. For example, Forbes' expectation of decreased energetic expense of his high-level responses is not shared by all (*contra* Halstead and O'Shea 1989). The axis along which Waddell scales his responses is geographic, whereas the others use social criteria. My model, for example, emphases social inclusivity as the measure of response cost. While not opposites, geographic distance is not necessarily the same as social distance.

RISK AVOIDANCE AND SOCIAL/CULTURAL CHANGE

What are the important relationships between culture change and environmental perturbations? How do problems of economic insecurity relate to changing human institutions? Some have proposed that increased severity simply and inextricably leads to the dissolution of society. There are many examples of social disruption with food scarcity. Laughlin, for example, in a study of the So of Uganda, demonstrated, "a tendency toward reduction of reciprocal exchanges of the general type in response to decreased basic resource availability" (Laughlin 1974:391). Those who focus on the psychological and physiological effects of starvation seem to view food shortages as leading inextricably to social disruption (e.g., Jelliffe and Jelliffe 1971). For example, Keyes, a pioneer in the study of the human physiological responses to starvation, quaintly and incorrectly noted that, "...in both primitive and civilized societies affected by famine the social ties are loosed, the usual social amenities and graces are dropped, and wives and children abandoned, and homes left" (Keyes et al. 1950:785). Not only is this view sexist, but it treats human behaviors as rather unimaginative and mechanistic. Keyes' view is not unique. Probably a majority of the readers of Turnbull's (1972, 1978) descriptions of the Ik society would come to the same conclusion as Keyes.

There is an alternative view. Other theories suggest that increased economic insecurity can result in or foster greater social cohesion. For example, a number of current archaeological models view the development of relationships beyond the household or kin group as the result of the need to buffer perturbation by expanding resource access through social networks, trade, and/or redistribution. Or stated differently, more inclusive social relations that allow wider access to food resources provide a higher probability of physical and social reproduction than other strategies.

Which perspective is correct? Is increased economic uncertainty related to, if not a cause of, social disintegration instead of social integration? The answer is, I propose, yes and yes. In order to resolve this paradox, it is important to recognize differing levels of food shortage severity and the ability of human populations to anticipate problems (e.g., Bennett 1976).

It might be useful to differentiate between what for the lack of better terms I will call *catastrophic* and *impinging* shortages; the former are rarer, and the latter are more common. Catastrophic events are those whose severity extends beyond the capacity of the population/cultural system to absorb the stress through "normal" responses. There can be both spatial and temporal components of catastrophic shortages; they can be severe in intensity, widespread in extent, and interminable in length. These events can lead to the social disintegration so often recorded, especially when combined with epidemics, as such events often are. Catastrophic shortages can set conditions for radical social/culture transformations through such processes as migration or ethnogenesis, the recombination of people into new social groups (Moore 1992; Sturtvevant 1971). The idea of "punctuated equilibria" in paleontology and evolutionary ecology is related to the concept of catastrophic perturbations.

Impinging shortages are those handled reasonably well through "normal" responses. Scholars who do atemporal studies can easily assume that such events have little or no effect on social/cultural change other than the development of low-level responses like storage or polyculture. They see what appears to be stasis. Impinging events, especially when combined with increasing population density or a decreasing resource base, can lead to social change, because these events can shift relationships ever so slightly. For example, a change in multiple field locations could well increase or change land ownership and alter nodes of interaction between social groups. The kinds of social change related to impinging events are analogous to incremental change in biological populations. These changes are more a trajectory than a transformation.

Two factors, other than the stress severity, are especially important to catastrophic and impinging events. One is the time lag of the perturbation for humans to consciously or unconsciously develop solutions. The second is the novelty of the problem. The greater the lag time available and the more familiar a population is with the problem, the more likely that the stress will be less disruptive. Human groups can and do anticipate, as the voluminous literature on hazard and risk perception attests, although it is also clear than anticipation/perception is rarely perfect. A short-lived school of natural hazards research in anthropology in the mid-1950s recognized the key role of anticipation when the following generalization was proposed: "social disorganization and stress are the greater as a) the disaster force is more rapid, b) the period of forewarning briefer, c) the disaster agent less well known and less clearly perceived" (Demerath 1957:2).

I have suggested that all types of responses, whether the most mundane or most radical, can relate to culture change, depending on the type of food shortage experienced and the context of the stress. The fact that a response is socially more inclusive, involves greater disruption of existing cultural relationships, or is energetically more expensive does not necessarily make it more likely to result in culture change. However, lower-level responses may be more likely related to incremental changes, while more catastrophic changes will associate with more disruptive social change.

SUMMARY

Given the evolutionary significance of food shortages, it should not surprise us that there are numerous strategies to ameliorate food stress. These range from low-level, household-oriented actions such as conservation of foodstuffs to more socially inclusive, communal actions that involve large numbers of people and may be coordinated by a small segment of the population. As expected, anthropologists place greatest emphasis on social responses. It is argued that the description of risk-reduction strategies in the archaeological record itself has practical value. The examples of famine food use in the prehistoric and ethnohistoric North American Southwest provide a cautionary tale about the suitability of ethnographic analogy for the study of prehistoric food stress responses.

Numerous scholars, coming from different disciplines and perspectives, have proposed that there is a pattern of responses to food shortages. While it may not be possible to predict what specific responses would or will be used to reduce the effects of food shortages, it may be possible to develop a generally applicable model of the order of responses to the scarcity of a range of resources. Less costly and more reversible responses will be used before more costly and less reversible responses; the magnitude of the response should match the magnitude of the problem. Several proponents of this approach, including myself, have argued that social relationships are an especially important response. My model explicitly suggests that response cost might be measured by the social inclusivity of the response.

It was further argued that contrary to what might be an intuitive view, all types of responses can affect, and be affected by, culture change. The fact that a response is socially more inclusive, involves greater disruption of existing cultural relationships, or is energetically more expensive does not necessarily make it more likely to result in culture change.

Researchers have noted that food stress is linked to both social dissolution and social integration. It is suggested here that relatively minor stresses are more likely to be related to incremental changes in society and culture, whereas catastrophic food shortages are more linked to major transformational changes such as ethnogenesis, the shifting hegemonies in areas of competing states, regional abandonment, and rapid major changes in the organization of society and subsistence complexes.

ACKNOWLEDGMENTS

Primary acknowledgements go to the participants of the Santa Fe Institute seminar and to the seminar's organizers, Joseph Tainter, George Gumerman, Linda Cordell, and Murray Gell-Mann. The staff of the Santa Fe provided an excellent setting. Patricia Gilman and Suzanne Fish aided in manuscript preparation.

REFERENCES

Akong'a, Joshua

1988 Drought and Famine Management in Kitui District, Kenya. In *Anthropology of Development and Change in East Africa*, edited by D. Brokensha and P. Little, pp. 99–120. Westview Press, Boulder.

Ali, Mehtabunsia

1984 Women in Famine: The Paradox of Status in India. In *Famine as a Geographic Phenomenon*, edited by B. Curry and G. Hugo, pp. 113–133. D. Reidel, Dordrecht.

Bennett, John W.

1976 *The Ecological Transition*. Pergamon Press, New York.

Berry, Michael S.

1982 *Time, Space, and Transition in Anasazi Prehistory*. University of Utah Press, Salt Lake City.

Bradfield, Maitland

1971 *The Changing Pattern of Hopi Agriculture*. Royal Anthropological Institute of Great Britain and Ireland Occasional Papers No. 30. London.

Braun, David P., and Stephen Plog
1982 Evolution of Tribal Social Networks: Theory and Prehistoric North American Evidence. *American Antiquity* 47:504–525.

Bunzel, Ruth
1932 Zuni Ritual Poetry. *Annual Report of the Bureau of American Ethnology, 1929–1930* 47:611–863.

Cashdan, Elizabeth (editor)
1990 *Risk and Uncertainty in Tribal and Peasant Economies.* Westview Press, Boulder.

Collier, George A.
1975 *Fields of the Tzotil.* University of Texas Press, Austin.

Colson, Elizabeth
1980 In Good Years and In Bad: Food Strategies of Self-Reliant Societies. *Journal of Anthropological Research* 35:1–029.

Cove, John N.
1978 Survival or Extinction: Reflections on the Problem of Famine in Tsimishian and Kaguru Mythology. In *Extinction and Survival in Human Populations*, edited by C. Laughlin and I. Brady, pp. 231–244. Columbia University Press, New York.

Dean, Jeffrey S., Robert C. Euler, George J. Gumerman, Fred Plog, Richard Hevly, and Thor N.V. Karlstrom
1985 Human Behavior, Demography, and Paleoenvironment on the Colorado Plateau. *American Antiquity* 50:537–554.

Demerath, Nicholas J.
1957 Some General Propositions: An Interpretive Framework. *Human Organization* 16:1–2.

Di Peso, Charles C.
1974 *Casas Grandes: A Fallen Trading Center of the Gran Chichimeca.* Northland Press, Flagstaff, Arizona.

Euler, Robert C., George J. Gumerman, Thor N. V. Karlstrom, Jeffrey S. Dean, and Richard H. Hevly
1979 The Colorado Plateaus: Cultural Dynamics and Paleoenvironment. *Science* 205:1089–1101.

Fish, Paul R., and Suzanne K. Fish
1989 Hohokam Warfare from a Regional Perspective. In *Cultures in Conflict: Current Archaeological Perspectives*, edited by D. Tkaczuk and B. Vivian, pp. 112–129. Proceedings of the Twentieth Annual Chacmool Conference. University of Calgary, Alberta.

Fish, Suzanne K., Paul R. Fish, and John Madsen
1985 Prehistoric Agave Cultivation in Southern Arizona. *Desert Plants* 7:107–113.

Forbes, H.
1989 Of Grandfathers and Grand Theories: Hierarchised Ordering of Responses to Hazard in a Greek Rural Community. In *Bad Year Economics: Cultural*

Responses to Risk and Uncertainty, edited by P. Halstead and J. O'Shea, pp. 87–97. Cambridge University Press, Cambridge.

Ford, Richard I.

1972 *An Ecological Analysis Involving the Population of San Juan Pueblo, New Mexico.* Ph.D. dissertation, Department of Anthropology, University of Michigan, Ann Arbor. University Microfilms, Ann Arbor.

1981 Gardening and Farming Before A.D. 1000: Patterns of Prehistoric Plant Cultivation North of Mexico. *Journal of Ethnobiology* 1:6–27.

1985 *Prehistoric Food Production in North America.* Anthropological Papers No. 75. Museum of Anthropology, University of Michigan, Ann Arbor.

1988 Personal communication.

Galt, Anthony H., and Janice W. Galt

1979 Peasant Use of Some Wild Plants on the Island of Pantelleria, Sicily. *Economic Botany* 32:20–26.

Gerald, Rex E.

1976 *Drought Correlated Changes in Two Prehistoric Pueblo Communities in Southeastern Arizona.* Ph.D. dissertation, Department of Anthropology, University of Chicago.

Gumerman, George J. (editor)

1988 *The Anasazi in a Changing Environment.* Cambridge University Press, Cambridge.

Haas, Jonathan

1989 The Evolution of the Kayenta Regional System. In *The Sociopolitical Structure of Prehistoric Southwestern Societies*, edited by S. Upham, K. Lightfoot, and R. Jewett, pp. 491–508. Westview Press, Boulder.

Halstead, Paul, and John O'Shea

1989 Introduction. In *Bad Year Economics: Cultural Responses to Risk and Uncertainty*, edited by P. Halstead and J. O'Shea, pp. 1–5. Cambridge University Press, Cambridge.

Hardin, G.

1968 The Tragedy of the Commons. *Science* 162:1243–1248.

Hardin, G., and J. Baden

1977 *Managing the Commons.* W. H. Freeman, San Francisco.

Horowitz, Michael M.

1976 *Colloquium on the Effects of Drought on the Productive Strategies of Sudano-Sahelian Herdsman and Farmers.* Institute for Development Anthropology, Binghamton, New York.

Jelliffe, Derrick B., and E. F. P. Jelliffe

1971 The Effect of Starvation on the Function of the Family and of Society. In *Famine: A Symposium Dealing With Nutrition and Relief Operations in Times of Disaster*, edited by G. Blix, pp. 54–62. Swedish Nutrition Foundation, Stockholm.

Keyes, A., J. Brozek, A. Henschel, O. Michelson, and H. L. Tayler
1950 *The Biology of Human Starvation.* University of Minnesota Press, Minneapolis.

Kirkby, Anne V. T.
1973 *The Use of Land and Water Resources in the Past and Present Valley of Oaxaca, Mexico.* Memoir No. 5. Museum of Anthropology, University of Michigan, Ann Arbor.
1978 Individual and Community Responses to Rain Fall Variability in Oaxaca, Mexico. In *Natural Hazards: Local, National, and Global,* edited by G. White, pp. 119–128. Oxford University Press, Oxford.

Laughlin, Charles D., Jr.
1974 Deprivation and Reciprocity. *Man* 9:380–396.

Lightfoot, Kent G.
1979 Food Redistribution Among Prehistoric Pueblo Groups. *The Kiva* 44:319–340.

Maclachlan, Morgan D.
1983 *Why They Did Not Starve: Biocultural Adaptations in a Southern Indian Village.* Institute for the Study of Human Issues, Philadelphia.

Marcus, Joyce
1982 The Plant World of the Sixteenth- and Seventeenth-Century Lowland Maya. In *Maya Subsistence: Studies in Memory of Dennis E. Puleston,* edited by K. Flannery, pp. 239–273. Academic Press, New York.

Marten, Gerald G., and Daniel M. Saltman
1986 The Human Ecological Perspective. In *Traditional Agriculture in Southeast Asia,* edited by G. Marten, pp. 20–54. Westview Press, Boulder.

Matson, R. J.
1992 *The Origins of Southwestern Agriculture.* University of Arizona Press, Tucson.

McCay, Bonnie J., and James M. Acheson
1987 *The Question of the Commons: The Culture and Ecology of Communal Resources.* University of Arizona Press, Tucson.

McCloskey, D. M.
1976 English Open Field as Behavior Toward Risk. *Research in Economic History* 1:144–170.

McLoughlin, Peter F. M.
1970 *African Food Production Systems: Cases and Theory.* Johns Hopkins University Press, Baltimore.

Minnis, Paul E.
1981 *Economic and Organizational Responses to Food Stress by Non-stratified Societies: A Case from the Prehistoric Southwest.* Ph.D. dissertation, Department of Anthropology, University of Michigan, Ann Arbor. University Microfilms, Ann Arbor.
1985a *Social Adaptation to Food Stress: A Prehistoric Southwestern Example.* University of Chicago Press.

1985b Domesticating People and Plants in the Greater Southwest. In *Prehistoric Food Production in North America*, edited by R. Ford, pp. 309–339. Anthropological Papers No. 75. Museum of Anthropology, University of Michigan, Ann Arbor.

1991 Famine Foods of the Northern American Desert Borderlands in Historical Perspective. *Journal of Ethnobiology* 11:231–257.

Moore, John H.

1992 An Ethnogenetic Critique of Cladistic Theory. Paper presented at the 158th Annual Meeting of the American Association for the Advancement of Science, Chicago.

Noy-Meir, I.

1985 Desert Ecosystem Structure and Function. In *Ecosystems of the World 12A: Hot Deserts and Arid Shrublands*, edited by M. Evenari, I. Noy-Meir, and D. Goodall, pp. 93–104. Elsevier, Amsterdam.

Page, Gordon B.

1940 Hopi Agricultural Notes. Mimeographed Ms. in author's possession.

Rappoport, Roy

1979 *Ecology, Meaning, and Religion.* North Atlantic Book, Richmond, CA.

Rautman, Alison

1990 *The Environmental Context of Decision-making: Coping Strategies Among Prehistoric Cultivators of Central New Mexico.* Ph.D. dissertation, Department of Anthropology, University of Michigan, Ann Arbor. University Microfilms, Ann Arbor.

Read, Bernard E.

1946 *Famine Foods Listed in the Chui Huang Pen Ts'ao.* (Reprinted edition in 1977). Southern Materials Center, Taipei.

Reid, J. Jefferson

1978 Responses to Stress at Grasshopper Pueblo, Arizona. In *Discovering Past Behavior: Experiments in the Archaeology of the American Southwest*, edited by P. Grebinger, pp. 195–214. Gordon and Breach, New York.

Reining, Conrad C.

1970 Zande Subsistence and Food Production. In *African Food Production Systems: Cases and Theory*, edited by P. McLoughlin, pp. 125–164. Johns Hopkins University Press, Baltimore.

Robbins, Wilfred W., John P. Harrington, and Barbara Freire-Marreco

1916 *Ethnobotany of the Tewa.* Bulletin 55. Bureau of American Ethnology.

Roys, Ralph L.

1967 *The Book of Chilam Balam of Chumayel.* University of Oklahoma Press, Norman.

Rudel, Thomas

1980 Social Responses to Commodity Shortages: The 1973–1974 Gasoline Crisis. *Human Ecology* 8:193–212.

Slobodkin, Lawrence B.

1964 The Strategy of Evolution. *American Scientist* 52:342–357.

1968 Toward a Predictive Model of Evolution. In *Population Biology and Evolution*, edited by R. Lewontin, pp. 187–205. Syracuse University Press, Syracuse.

Slobodkin, Lawrence B., and Anatol Rapoport
1974 An Optimal Strategy of Evolution. *Quarterly Review of Biology* 49:181–200.

Spielmann, Katherine
1982 *Intersocietal Food Acquisition Among Egalitarian Societies: An Ecological Study of Plains-Pueblo Interaction in the American Southwest.* Ph.D. dissertation, Department of Anthropology, University of Michigan, Ann Arbor. University Microfilms, Ann Arbor.

Stanton, Glen, D. R. Burnham, and John Carter, Jr.
1932 *Corn Investigations in New Mexico.* Bulletin No. 260. Agricultural Experiment Station, New Mexico State University, Las Cruces.

Stephens, Alexander M.
1936 *Hopi Journal.* Contribution to Anthropology No. 23 Columbia University, Manhattan.

Sturtevant, William C.
1971 Creek into Seminole. In *North American Indians in Historical Perspective*, edited by E. B. Leacock and N. O. Lurie, pp. 92–128. Random House, New York.

Tuan, Yi-Fu, Cyril E. Everard, Jerold G. Widdison, and Iven Bennett
1973 *The Climate of New Mexico.* New Mexico State Planning Office, Santa Fe.

Turnbull, Colin M.
1972 *The Mountain People.* Simon and Schuster, New York.
1978 Rethinking the Ik: A Functional Non-social System. In *Extinction and Survival in Human Populations*, edited by C. McLaughlin and I. Brady, pp. 49–75. Columbia University Press, New York.

Waddell, Eric
1972 *The Mound Builders: Agricultural Practices, Environment, and Society in the Central Highlands of New Guinea.* University of Washington Press, Seattle.
1975 How the Enga Cope with Frost: Responses to Climatic Perturbations in the Central Highlands of New Guinea. *Human Ecology* 5:69–76.

Watt, M.
1988 Coping with the Market: Uncertainty and Food Security Among the Hausa Peasants. In *Coping with Uncertainty in Food Supply*, edited by I. de Garine and G. A. Harrison, pp. 260–289. Clarendon Press, Oxford.

Whiting, Alfred F.
1939 *Ethnobotany of the Hopi*. Bulletin no. 15. Museum of Northern Arizona, Flagstaff.

Wilcox, David
1989 Hohokam Warfare. In *Cultures in Conflict: Current Archaeological Perspectives*, edited by D. Tkaczuk and B. Vivian, pp. 163–172. Proceedings of the Twentieth Annual Chacmool Conference. University of Calgary, Alberta.

Wills, Wirt H., III
1988 *Early Prehistoric Agriculture in the American Southwest.* School of American Research Press, Santa Fe.

Katherine A. Spielmann and Eric A. Angstadt-Leto
Department of Anthropology, Arizona State University

Hunting, Gathering, and Health in the Prehistoric Southwest

INTRODUCTION

Over the past several decades, a number of individual and a few synthetic analyses have been undertaken concerning the nutritional health of prehistoric Southwestern agricultural populations (e.g., El-Najjar et al. 1976; Martin et al. 1991; Nelson et al. 1992; Palkovich 1980; Stodder 1990). Taken together these studies document a fair degree of variability in health across the Southwest, particularly as reflected in the incidence of porotic hyperostosis, an osteological condition indicative of iron deficiency anemia (Stuart-Macadam 1987). The incidence and severity of anemia are the product of several different variables, among them infection, parasite load, and diet. Systematic studies of each of these variables are necessary before we can understand the underlying causes of variation in Southwestern agriculturalists' health.

In this chapter, we have chosen to investigate the relationship between diet and health. In particular, we focus on variability in prehistoric access to meat and wild plant foods across the Southwest. First we consider the nutritional consequences of meat-poor diets. We next document a great deal of variability in faunal assemblage composition, which likely reflects variability in access to meat among prehistoric Southwestern populations. We then move on to an analysis of archaeological data relating to three alternative means of dealing with shortfalls in local meat availability: trade for meat, turkey husbandry, and harvest of plants containing nutrients found in meat. In assessing the latter strategy, we focus on beans and on wild plants high in iron and vitamin C.

The sites included in this analysis are those of aggregated populations, as several factors are likely to make access to meat and wild plants more problematic for larger than for smaller, highly dispersed populations. Among these factors are population density, which may be too high to rely on wild plants locally, overhunting (Speth and Scott 1989; Szuter and Bayham 1989; Szuter 1992), and environmental degradation (Szuter 1992). Aggregated populations may also compete with one another for access to more distant sources of game. Thus, the severity of nutritional stress is likely to be greater in aggregated populations, possibly making it more detectable archaeologically.

Botanical (Raymer and Minnis 1992; Minnis 1989) and bone chemistry (Martin et al. 1991:67-76; Matson and Chisholm 1991; Spielmann et al. 1990; Spielmann and Schoeninger 1992) data have demonstrated that Southwestern farming populations were heavily dependent on maize for the bulk of their diet. For reasons elaborated upon below, a maize-dominated diet is deficient not only in high-quality protein but also in essential minerals, such as iron, and vitamins, such as B12. Such deficiencies can have a variety of consequences, including anemia, neurological damage, and increased incidence of infectious disease. Thus, our expectation is that Southwestern populations would have had to consume animal meat and/or certain plants regularly in order to maintain nutritional health. Restrictions in access to these foods are expected to result in lower nutritional health, which may be evidenced skeletally.

Our emphasis here is as much on the vitamins and minerals available in meat as it is on the protein value of meat. We believe that it is important to view meat not only as a source of protein, as is generally the case in most archaeologically based analyses of nutrition, but also as a source of other essential nutrients. One corollary of this emphasis is that while Southwestern archaeologists' traditional fascination with corn production is important for understanding prehistoric diets, information concerning access to meat is equally critical.

NUTRITIONAL CONSIDERATIONS

IRON

Animals and plants provide two distinctly different sources of iron, heme and non-heme, in the human diet. Iron is found in heme, red pigment in the blood, and in this form is more easily absorbed by humans. Ten to twenty-five percent of heme iron is absorbed by the body. Obviously, heme iron is only available from animal resources. Non-heme iron, which is available in plants, is low in bioavailability, the degree to which it is absorbed by the body. Only one to two percent of non-heme iron is absorbed, if not eaten with other foods (Hallberg 1981).

The absorption of non-heme iron is significantly affected by the iron status of the individual and by characteristics of the meal in which the iron is consumed. Individuals who are anemic will absorb more non-heme iron than non-anemic individuals who eat the same meal. Both meat and vitamin C (ascorbic acid) greatly enhance the absorption of non-heme iron, with meat increasing absorption of non-heme iron from 1–2% up to 25%. When meat is added to a meal of corn, the absorption of iron increases three-fold (Layrisse et al. 1968). Vitamin C also acts as a facilitator of non-heme iron uptake, often increasing iron absorption by a factor of four (Hallberg 1981). Meal components that inhibit iron absorption are phytates and tannins (Banerji et al. 1968; Baynes and Bothwell 1990; Dwyer 1991; Hallberg 1981; Kuhn et al. 1968). Phytates are found in grains, including corn.

Research concerning the bioavailability of iron in different meals and in different dietary regimes indicates that there is great variation in the iron availability in meals of the same caloric content (Hallberg 1981), and among diets that characterize different populations worldwide (Baynes and Bothwell 1990). Monotonous diets high in cereals, roots, or tubers with little meat, fish, or vitamin C provide around 0.7 mg. of iron per day. This quantity of iron is insufficient for the needs of women, especially pregnant or lactating women, children, and many men (Baynes and Bothwell 1990). Such diets are typical of populations in many developing countries, and may have been characteristic of some prehistoric Southwestern populations.

Diets that rely on the above sources of carbohydrate, but include some meat or fish and vitamin C, provide on average about 1.4 mg. of iron per day. This quantity meets the needs of adult men, and women who are not pregnant or lactating. Pregnant women require 2.2 mg. of iron per day simply to maintain iron at a nutritionally acceptable level. Many traditional diets do not provide this quantity of iron (Hallberg 1981). Moreover, adolescent needs may not be met by this kind of diet (Baynes and Bothwell 1990).

The biologic consequences of iron deficiency include impaired work capacity (Baynes and Bothwell 1990), and greater susceptibility to respiratory infections and diarrhea (Goodman 1994). Iron deficiency to the point of anemia can result in neurologic disfunction if it is experienced at an early age. This disfunction is not completely reversible by a better diet later in life (Baynes and Bothwell 1990).

VITAMIN B12

Vitamin B12 is available in substantial amounts only in animal products, meat or milk. Vegetarian diets, particularly vegan diets, which contain no animal products, have been documented to result in vitamin B12 deficiency. For example, a study of young children on macrobiotic diets documented vitamin B12 levels far below those in the control group, levels sufficiently low to have physiological consequences and to cause concern about the children's neurological development (Dagnelie et al. 1989b), since Vitamin B12 deficiency can cause neurological damage. Moreover, the children on macrobiotic diets exhibited problems in their growth and motor development (Dagnelie et al. 1989a, 1990; see also Dwyer et al. 1982). Modern vegetarians, particularly children and pregnant and lactating women, are strongly encouraged to supplement their diets with B12 (Dwyer 1991). In the prehistoric Southwest, any relatively meatless diet would also likely have resulted in vitamin B12 deficiency.

Given the potentially severe consequences of relatively meatless diets, we expect that meat was a critical commodity in the prehistoric Southwest. As documented below, however, faunal data suggest that meat from local game may have been insufficient in quantity to satisfy the nutritional needs of some aggregated Southwestern populations.

HUNTING IN THE PREHISTORIC SOUTHWEST

Rabbits and artiodactyls (primarily deer and antelope) dominate the faunal assemblages from aggregated prehistoric Southwestern sites. The large populations of such sites (often several hundred people or more) would have required sizeable quantities of game to meet basic nutritional requirements. Table 1 illustrates the calculation of the number of lagomorphs or artiodactyls that populations of 100 people and of 500 people would need per year using estimates of human protein needs, meat weights, and protein content. Unfortunately, the quantities of iron and vitamin B12 that humans need on an annual basis are not well established (e.g., Keene 1981). Here we use human protein needs to illustrate the quantities of game an aggregated population is likely to require.

Based on Food and Agriculture Organization/World Health Organization (1973) estimates for the protein needs of populations in developing countries, an average of approximately 20 g./person/day of high-quality protein is necessary to maintain nitrogen balance (Wetterstrom 1986:165). Rabbits (if the useable meat is averaged for cottontail and jackrabbit) produce roughly 0.94 kg. of meat, which at 21% protein content, provides 0.20 kg. of protein per animal. An "average" deer supplies 48.5 kg. of useable meat, and at 21% protein, 10.19 kg. protein per animal.

As the table illustrates, roughly 3650 rabbits or 72 deer would supply the annual protein needs of a population of 100. These figures rise to 18,250 rabbits or 360 deer with a population of 500.

Given that protein may also be derived from the consumption of small mammals and plants, however, we do not expect that all of the protein needs of prehistoric Southwestern populations were met by deer and rabbit meat. Thus, for example, if only one-half of the necessary quantity of protein were supplied by game, 500 people would require 9125 rabbits, or 180 deer per year (Table 1).

TABLE 1 Calculations of Protein Available from Artiodactyls and Lagomorphs.

I.	Available Protein Jackrabbit: 1.38 kg. useable meat Cottontail: .50 kg. Mule Deer: 48.5 kg. useable meat At 21% protein content: one Deer = 10.19 kg. protein one Rabbit = .20 kg. protein one Deer = 51 rabbits in protein content	Avg. .94 kg.
II.	Required Protein .02 kg./person/day on avg. × 100 people × 365 days = 730.0 kg. protein/population/year	
III.	Number of animals to meet protein needs per year At population of 100: Rabbits: 730/.20 = 3650 rabbits or Deer: 730/10.19 = 72 deer If 50% of protein needs are met by animal meat: Rabbits: 1825 or Deer: 36 At population of 500 and 50% of protein from animal meat: Rabbits: 9320 or Deer: 180	

These latter figures are sizeable, and it is likely that Southwestern populations varied in their ability annually to acquire faunal resources of this magnitude. Factors such as duration of site occupation, density of sites in an area, and the capacity of the environment to support faunal resources would affect the quantity of game available in the vicinity of a prehistoric site. In this chapter we seek to document that variability, and to assess alternative strategies to deal with insufficient access to game.

One means of comparing the relative availability of meat to prehistoric Southwestern populations is through the calculation of a ratio between artiodactyl and lagomorph bone found at archaeological sites (Spielmann 1991a; see also Szuter and Bayham 1989). This ratio is called the artiodactyl index, and is used as a measure of the relative accessibility of large game as a meat resource. In this analysis we have assumed that an index of large game accessibility reflects the accessibility of meat in general to aggregated populations. Because it would take roughly 50 rabbits to match the protein contribution of a single deer, we expect that there is a limit to the ability of rabbit meat to make up for a lack of availability of meat from large game.

It is important to point out, however, that it will be necessary to obtain independent data to evaluate whether the artiodactyl index does indeed reflect access to meat in general. Bone chemistry information will be especially critical in this assessment. In particular, trace element analyses measuring strontium and barium content in bone should provide information on the relative quantities of meat in the diets of different populations, as long as geographic variation in trace element distributions is controlled for. At this point, only three bone chemistry studies of aggregated Southwestern populations have been undertaken. Schoeninger and Spielmann have collected strontium and stable isotope data for Pecos (Spielmann et al. 1990) and Gran Quivira Pueblos (Spielmann and Schoeninger 1992). Ezzo (1992) has recently completed an extensive bone chemistry study of the Grasshopper burials. Comparisons among these data sets have yet to be undertaken, however.

ARTIODACTYL INDICES

In order to compare access to large game across the Southwest, Spielmann (1991a) computed an artiodactyl index for the faunal assemblages from a series of ten aggregated sites; several additional sites have been added to the original sample for the purposes of this analysis (see Table 2). The artiodactyl index as computed here is the ratio of artiodactyl NISP (Number of Identified Specimens; i.e., number of bone fragments identifiable to a particular taxon) to lagomorph NISP [see Bayham 1982].)

To make valid comparisons among NISPs from different sites, it is important to establish that the bone assemblages from these sites have not been differentially affected by processes that would lead to greater or lesser fragmentation of the species under analysis. For example, a number of taphonomic processes, including trampling and weathering as well as butchering patterns, can substantially affect the degree of bone fragmentation at a site, and thus the NISP. Taphonomic processes at the sites chosen for analysis (see Table 2), however, appear to have been similar. Based on the published reports on these sites it appears that the animal bone is in good condition, and that artiodactyls were butchered to a similar degree. For example, most long bones were broken for marrow, but there was generally not much further fragmentation. Moreover, the patterning in the artiodactyl ratios (see below) was strong enough that variation due to somewhat different taphonomic histories is not expected to obscure behaviorally meaningful trends in the index. Direct study of each collection, however, would be necessary to make a stronger case for the similarity in degree of fragmentation among the assemblages.

TABLE 2 Artiodactyl indices for aggregated towns.[1]

Site	NISP A	NISP L	A/L	Reference
Pueblo Colorado	1982	2151	.92	Thiel (1994)
Gran Quivira	2374	5987	.40	Spielmann, unpublished data
Rowe Pueblo[2]	960	800	1.20	Mick O'Hara (1987)
Arroyo Hondo	965	4885	.20	Lang and Harris (1984)
Pueblo Alto	908	10,708	.08	Akins (1987)
Chaco Villages[3]	225	3356	.07	Akins (1985)
Guadalupe Ruin	1354	1583	.86	Pippin (1987)
Galaz Ruin	147	714	.21	Anyon and LeBlanc (1984)
Mimbres Classic[4]	305	1319	.23	Nelson, unpublished data
Mimbres Salado[5]	450	173	2.60	Nelson, unpublished data
La Ciudad	158	1368	.12	Szuter (1989)
Los Colinas	227	3687	.06	Szuter (1989)

[1] NISP A = Number of identified artiodactyl specimens; NISP L = Number of identified lagomorph specimens.

[2] Excludes data from backhoe trenches since these sediments were not screened.

[3] Villages 627, 629, and 633.

[4] Sites LA676 and LA12076.

[5] Sites Z:1:78, Z:5:10, and LA12077.

In order to control for recovery biases, particularly against small game, only aggregated sites at which sediment was screened through 1/4-inch mesh were used in the analysis. Thus, biases due to recovery techniques should be similar among the sites. The contexts excavated generally included a mix of rooms, features, and middens at each site. Different excavations emphasized different contexts, however. For this pilot study, we are assuming that this difference in emphasis did not significantly affect the recovery of different species of animal or plant.

Outside the Hohokam area, there are relatively few published or readily accessible faunal analyses for aggregated settlements, particularly ones at which sediment was routinely screened during excavation. The sample discussed here includes four sites from the eastern border area of the Pueblo world, several sites in the Chacoan region, and several sites from the Mogollon highlands (see Figure 1 and Table 2). We have also included two sites in the Hohokam area: La Ciudad, a preclassic village, and Las Colinas, a pre-classic and classic site, both of which are on the Salt River.

As Table 2 and Figure 2 indicate, the artiodactyl index varies from 2.6 at Salado sites in the Mimbres area to .06 at Las Colinas. Values for the eastern-most pueblos, those nearest the plains and mountainous areas, and the Mogollon highlands are relatively high. In contrast, values for the western pueblos and Hohokam tend to be low. Given the expectation of some archaeologists that the large towns in Chaco Canyon served as feasting locations, and possibly elite residences, it is interesting to note the very low value of the artiodactyl index for Pueblo Alto, and the virtual identity of that index with the index for three small villages in the canyon (see also Akins 1985).

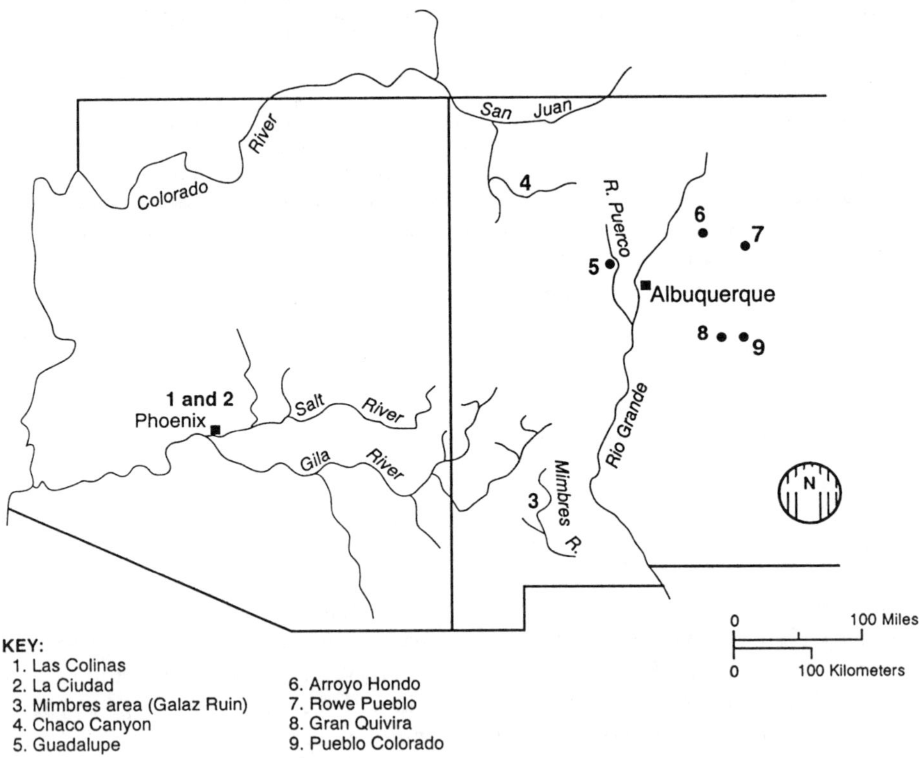

FIGURE 1 Map showing aggregated sites used in the analysis.

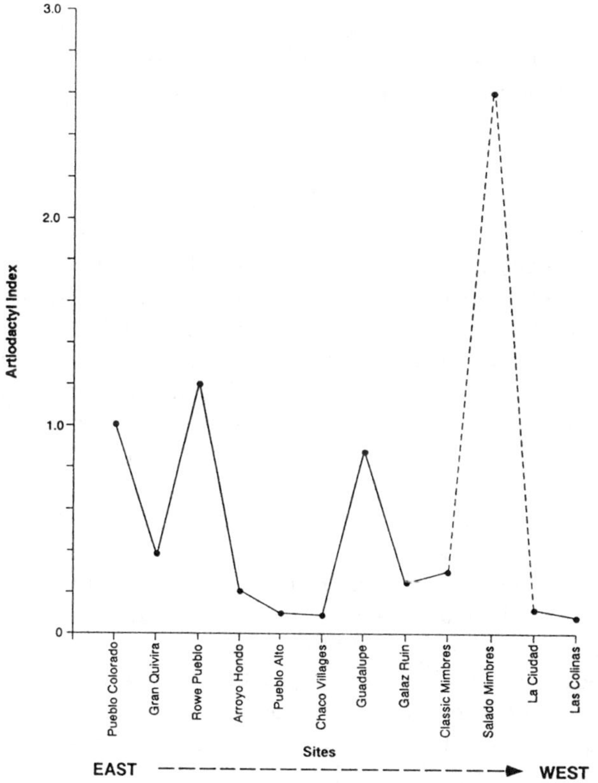

FIGURE 2 Artiodactyl indices for aggregated sites.

ALTERNATIVE STRATEGIES

If the artiodactyl index is reflective of access to large game, and access appears to have been low at many aggregated Southwestern sites, then one would expect that those populations with little direct access to large game might develop alternative strategies to acquire protein, minerals, and other nutrients. These strategies might include harvesting of very small game, trading for meat, raising turkeys (Speth and Scott 1989), greater emphasis on the cultivation of beans to complement the amino acids in corn, and greater focus on the gathering of wild plants that contain relatively high quantities of iron and other minerals that meat provides, but in which corn is deficient. Wild plants may also provide amino acids complementary to those in corn.

Szuter (1989, 1992) has cogently argued that rodent-sized game formed an important component of the lowland Hohokam diet, and may have made up for the lack of access to large game for Southwestern populations in general. Spielmann (1988) noted an increase in small game in Gran Quivira middens toward the end of the occupation, when Plains-Pueblo trade was disrupted by Spanish colonists and local large game may have been overhunted. Given that the strategy of small-game hunting is documented in existing literature, here we investigate other potential strategies: trade, domestic animals (turkey), beans, and wild plants.

TRADE

The sample of aggregated sites includes two groups, the lowland Sedentary and Classic Hohokam and the Chaco Anasazi, which many archaeologists believe reached a higher level of sociopolitical complexity than other prehistoric Southwestern societies. These populations maintained widespread trade networks through which sizeable quantities of goods such as shell and pottery moved. Trade has also been postulated as a means by which populations in both areas acquired meat from large game. The low artiodactyl indices for these areas may thus belie more abundant access to large mammal meat. In order to assess this possibility, data on artiodactyl abundance are needed from the areas that are thought to be the sources of traded large game.

In the Hohokam area, sites in the uplands north of the Salt/Gila basin are postulated to have provided meat to Hohokam centers to the south (Bayham and Hatch 1985; James 1987). In the Chaco area, Pueblo populations in the Chuska mountains, who supplied ceramics, chert, and wooden beams to Chaco, are thought to have been a significant supplier of meat from large game as well (Akins 1985:409). If these propositions are correct, then faunal assemblages at sites purported to be the sources of traded meat should exhibit high artiodactyl indices.

While data are lacking for the Chuska mountains, it is possible to evaluate, on a preliminary basis, data from upland locations in the Hohokam area (Figure 1, Table 3). Ash Creek, Miami Wash, and Anamax-Rosemont are upland locations which each contain several small Hohokam farming sites. The Ventana figures are from the top two levels at Ventana Cave, which Bayham (1982) has argued represent remains of Hohokam hunting camps.

Clearly the Ash Creek, Miami Wash, and Anamax-Rosemont sites display higher artiodactyl indices than the contemporaneous lowland Hohokam villages discussed previously, and their indices are not that different from Ventana Cave, particularly given that only 1/2″ mesh screen was used at Ventana. Szuter and Bayham (1989) have argued that higher artiodactyl indices for upland Hohokam sites reflect a greater emphasis on hunted meat in the upland diet, due to the greater abundance of large game in the uplands and the relatively small upland population sizes. In contrast, Bayham and Hatch (1985) and James (1987) have suggested that the higher artiodactyl indices for upland sites may be due to trade in meat from the uplands to the lowlands. Occupants of upland sites, in this scenario, are thought to have specialized to some degree in hunting large game, whose meat was then traded to people in lowland centers in return for goods manufactured there, and perhaps

TABLE 3 Artiodactyl indices for upland sites.[1]

Site	NISP A	NISP L	A/L	Reference
Ash Creek	72	267	.27	Szuter (1989:233)
Miami Wash	516	657	.79	Szuter (1989:234)
Anamax-Rosemont	785	1429	.55	Szuter (1989:233)
Ventana, Level 1	321	179	1.79	Szuter and Bayham (1989)
Ventana, Level 2	371	259	1.43	Szuter and Bayham (1989)

[1] For abbreviations, see Table 2.

for corn. At this point it is not possible to choose between these two scenarios. Bone chemistry data (see above), however, would assist in determining whether upland populations did indeed consume more meat than lowland populations.

Trade for meat between eastern border Pueblo farmers and plains hunter-gatherers has been well documented (see Spielmann 1991b and 1991c, and references therein). Such trade appears to have developed in the mid-1400s as evidenced by exchange data from the Plains (Spielmann 1983), and by the relative importance of bison in the artiodactyl assemblages from eastern border pueblos. For example, while bison comprises only 3% (31/960) of the artiodactyl NISP at Rowe, which was abandoned in the 1300s (Mick O'Hara 1987), at Gran Quivira 17% (399/2374) of the artiodactyl NISP is bison, with the majority of the bison bone from deposits dating after A.D. 1500 (Spielmann 1988).

TURKEY

Several years ago Speth and Scott (1989) suggested that turkey raising may have been one strategy that Southwestern populations pursued to increase their access to animal meat. Because turkey provides more than twice the amount of meat per animal than rabbit (2.3 kg./turkey; Wetterstrom 1986:173), turkey would seem to be a viable alternative to rabbit in compensating for shortfalls in access to large game. Obviously, however, there are costs in raising turkeys that are not taken into account by simply looking at the benefits (see Akins [1985] for a discussion of such costs).

The earliest domesticated turkeys from the San Juan Basin date to the seventh century A.D., and there is ample evidence of turkey domestication in the Rio Grande by A.D. 1200 (Lang and Harris 1984:93). Windes' (1987) review of turkeys at Pueblo Alto documents turkey husbandry there by the 1000s. Evidence for turkey husbandry includes remains of turkey pens, eggshells, and poults.

While most would not dispute the evidence of turkey raising in the Southwest, many have questioned whether turkeys were kept for their feathers, or whether they were a source of food. Windes' (1987) interesting analysis of turkey remains from Pueblo Alto documents a dramatic increase in the burning of turkey bone in the 1100s at that site, corresponding with a sizeable increase in the occurrence of turkey. On the basis of this evidence he proposes that turkeys were used for feathers at Pueblo Alto in the 1000s and were consumed as food in the 1100s. Lang and Harris (1984) assume that the turkeys at Arroyo Hondo were consumed throughout the occupational sequence.

In contrast, McKusick (1981) has maintained that turkeys at Gran Quivira were raised solely for their feathers. Her argument rests on the age profile of the turkeys, the lack of burned turkey bone, and the lack of cut marks on the bone. She argues that turkeys were killed at too old an age (over two years) to be palatable, although she does note that there is a relatively high rate of juvenile male turkey mortality in the sample. With regard to juvenile male mortality, Peterson (1989) points out that because one gobbler can service several female turkeys, many juvenile males at Gran Quivira may have been culled from the flock since they would not have been necessary for flock production and maintenance. Presumably these younger individuals would have been palatable and could have been eaten.

In summary, there were apparently both dietary and nondietary uses of turkey. Our expectation is that aggregated populations had a fairly fixed need for feathers, and that relatively high proportions of turkey bone at a site would signify use above these needs, i.e., meat consumption. Burning data, however, also appear to be important in identifying the consumption of turkey meat.

In order to evaluate the possibility that turkey compensated for a shortfall in large mammal protein, NISPs for turkey were compiled and a "turkey index" was calculated by dividing turkey NISP by lagomorph NISP for each of the twelve sites in the sample. This index was then compared with the artiodactyl index. It was expected that if turkey did indeed make up for a lack of large game availability at some sites, then the turkey index should vary inversely with the artiodactyl index. Statistics were not computed in comparing these indices, however, due to the unsuitability of NISP data for statistical analysis.

Counts of turkey bone include only those elements identified as turkey in the faunal reports. Although elements in "large bird" categories were likely to be turkey as well, not all reports included such a category, and thus we were limited by the nature of the published data. While turkey frequency may thus be underestimated at some sites, we do not expect that the trends noted in the data would be changed if large bird elements were added to the sample.

Table 4 provides NISPs for turkey and the calculation of the turkey index. Sites are arranged in descending order based on their artiodactyl index. In this table sites with indices above zero fall into two categories, those with very low values (<.10) and those with higher values (>.20). Our suspicion is that the low indices reflect a low but constant demand for turkey feathers, while the higher indices reflect meat consumption. This proposition is borne out by data from sites in Chaco Canyon discussed below.

Figure 3 graphs both the turkey and artiodactyl indices for each site. This figure clearly shows that there is no correlation between the turkey and artiodactyl indices. The group of sites with the highest artiodactyl index (Mimbres Salado), has one of the lowest turkey indices, while the site with the second-highest artiodactyl index has the highest turkey index (though turkey at Rowe do not appear to have been raised at the site [Mick O'Hara 1987:6]). Many sites with low artiodactyl indices also have low turkey indices.

It is likely that several factors other than access to large game are affecting the turkey index. First, some areas, such as the Sonoran desert, may not have been conducive to turkey husbandry, perhaps because turkey are adapted to higher

TABLE 4 Turkey indices for SW sites.[1]

Site	NISP T	NISP L	T/L
Mimbres Salado[2]	3	173	.02
Rowe Pueblo	249	800	.31[3]
Pueblo Colorado	43	2151	.02
Guadalupe Ruin	345	1583	.22
Gran Quivira	198	5987	.03
Classic Mimbres[4]	22	1319	.02
Galaz	0	714	0
Arroyo Hondo	1134	4885	.23
La Ciudad	0	1368	0
Pueblo Alto	987	10798	.09
Chaco Villages	802	3356	.24
Las Colinas	0	3687	0

[1] Sites are arranged in decreasing order based on the artiodactyl index value. NISP T = Number of identified turkey specimens. NISP L = Number of identified lagomorph specimens.

[2] Sites Z:1:78, Z:5:10, and LA12077.

[3] No eggshell or juveniles found; likely these were not kept at the site (Mike O'Hara 1987:6).

[4] Sites LA676 or LA12076.

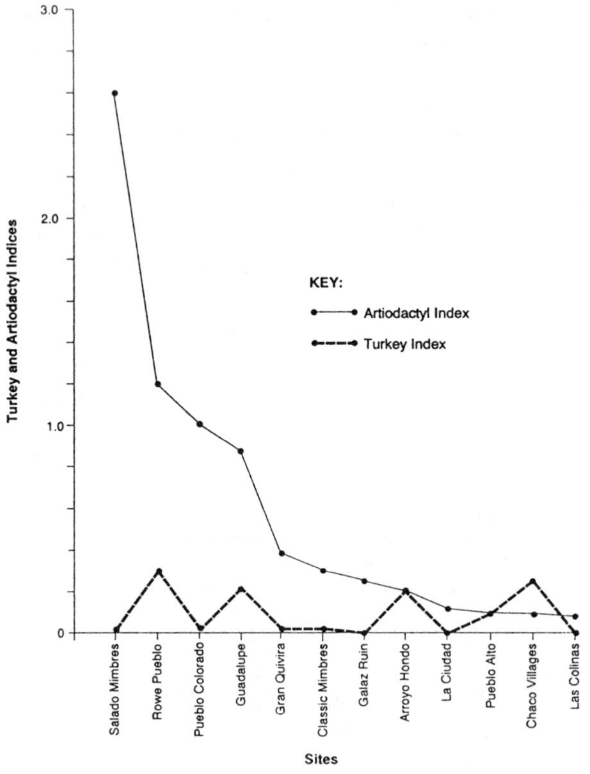

FIGURE 3 Turkey and artiodactyl indices for aggregated sites.

elevations and cooler temperatures. The Hohokam sites in the sample lacked turkey bone. Moreover, Szuter (1989:189) states that in general turkey are rare in Sonoran Desert sites. Second, the taboos against turkey consumption mentioned by Spanish chroniclers may have been operative in some regions such as the Salinas area (Gran Quivira) and among the Mimbres. The low turkey indices in both these areas may reflect use of turkey feathers.

There are data from Chaco Canyon, however, that suggest an increase in emphasis on turkey husbandry as large game became less available. In order to detect a change in turkey husbandry, we divided the Chacoan assemblage temporally.

Table 5 provides the artiodactyl and turkey indices for the Chaco villages and Pueblo Alto over time. The Chaco village data have been divided between sites 627 and 629, which contain deposits dating from between A.D. 850 and 1200, and site 633, with deposits dating from roughly A.D. 1220 to 1250 (Akins 1985). The Pueblo Alto data follow Akins' (1987:Table 8.144, p. 624) divisions among the Red Mesa, Gallup, and Late Mix deposits, which date roughly A.D. 920–1020, 1020–1120, and 1120–1220, respectively. Noteworthy is the fact that the turkey index is fairly low until the late phases of occupation at both the town and village sites.

Also noteworthy is the relatively low and static artiodactyl index at Pueblo Alto over time. As discussed above, the low index may mask the importance of imported, but relatively boneless, meat from the Chuska Mountains. Clearly if there were a shortfall in large game over time at Alto, lagomorphs were not being taken in increasing numbers to make up for it.

Akins (1985) has attributed the increase in turkey frequency in Chaco Canyon to the availability of abandoned cornfields for turkeys to forage in as the canyon became depopulated. An alternative explanation is that as the Chaco system collapsed, the system for provisioning the canyon with large game from the Chuskas also collapsed. In response, the remaining Chaco Canyon population, in both the villages and the towns, turned to an alternative source of animal meat, the domestic turkey.

Trends in Rio Grande turkey usage may parallel those in Chaco Canyon, only at a later date given that populations did not begin to aggregate in the Rio Grande until ca. A.D. 1300. Lang and Harris (1984:100) note the apparent increase in importance of turkey during the late prehistoric period. Turkey bone percentages range between 9% and 15% for sites in the Rio Grande Valley that date to the 1300s, while sites dating to the 1400s and 1500s contain percentages of 23 to 29%. In contrast, further to the east preliminary analysis of the turkey from Gran Quivira (Peterson 1989) suggests no significant change in the emphasis on turkey over time at that site, despite its growth in size between the 1300s and 1400s. McKusick (1981) may thus be correct in suggesting that turkeys were raised at Gran Quivira predominantly for feathers.

In sum, turkey husbandry is not a strategy used uniformly across the Southwest to allay shortfalls in access to large game. Where temporally separable faunal data are available for the Anasazi area, however, the frequency of turkey increases over time; perhaps this was in response to overhunting or to the demise of trading systems that originally provisioned some of these populations with large game.

TABLE 5 Turkey and artiodactyl indices for Chaco sites.[1]

Site	NISP T	NISP A	NISP L	T/L	A/L
Chaco 627/629	118	219	1962	.06	.11
Chaco 633	684	6	1282	.53	.00
Alto Red Mesa	3	101	1703	0	.06
Alto Gallup	68	447	5653	.01	.08
Alto Late Mix	878	284	2906	.30	.10

[1] For abbreviations, see Tables 2 and 4.

The lack of turkey in the Sonoran Desert sites would seem to reinforce the importance of rodent-sized game to the subsistence of Hohokam populations. One important avenue of further inquiry would be to compare the relative emphasis on very small game between Colorado Plateau and Sonoran Desert populations. Our expectation is that an emphasis on very small game should be negatively correlated with an emphasis on turkey husbandry.

PLANT RESOURCES

If trade and/or turkey husbandry were strategies that were either not available to some Southwestern populations, or were insufficient to alleviate a shortfall in access to meat from large game, an emphasis on the cultivation or harvesting of some plant species constitutes a third potential strategy for acquiring nutrients provided by meat. While plants do not provide vitamin B12, they do contain other critical nutrients found in meat. In this section we consider one such nutrient, iron, and assess the degree to which reliance on plants high in iron correlates with lack of access to large game. In addition, we investigate the prehistoric use of plants high in vitamin C, as this nutrient enhances the absorption of non-heme iron (see above).

Before undertaking a comparative analysis of wild plant use at different Southwestern sites, it was necessary to assess the degree of comparability of the macrobotanical assemblages from these sites. Information on the macrobotanical recovery procedure used is scarce for sites excavated earlier than about 1980. Generally, however, an initial sample of 2–4 liters of sediment was processed through a flotation device, and the plant remains were then sorted and analyzed. With regard to the plant identifications, we were heartened by the fact that only two researchers (Minnis and Toll) were involved in the analysis of 80% of the sites considered here (Table 6). This situation reduces the problem of interobserver differences in analytical techniques. In addition, we use ubiquity data to compare the utilization of plants at the sites. The ubiquity of a taxon is calculated by dividing the number of flotation samples the taxon occurred in by the total number of flotation samples analyzed. This presence-absence measure helps to minimize differences among sampling procedures. Table 6 shows the number of macrobotanical samples analyzed for each site and the researchers involved.

The two Hohokam sites used in the faunal analysis, La Ciudad and Las Colinas, were eliminated from this portion of the study. Because these sites are situated in the Sonoran Desert, their floral assemblages differ significantly from those recovered from the Mogollon and Anasazi sites. Since the same species could not be compared at the Hohokam sites we do not consider them here.

TABLE 6 Macrobotanical samples and ethnobotanists.

Site	Sample Size	Analyst
Mimbres Salado	43	Paul Minnis (1986)
Rowe Pueblo	15	Mollie Toll (1981)
Pueblo Colorado	26	Michael Diehl (1990) [with Paul Minnis]
Guadalupe Ruin	18	Vorsila Bohrer & Karen Adams (Pippin 1987)
Gran Quivira	50	Raymer and Minnis (1992)
Mimbres Classic	35	Paul Minnis (1985)
Galaz Ruin	16	Paul Minnis (1984)
Arroyo Hondo	174	Wilma Wetterstrom (1986)
Pueblo Alto	124	Mollie Toll (1987)
Chaco villages	164	Mollie Toll (1985)

The species of plants considered in our analysis were chosen using two criteria: availability of nutritional information on iron and vitamin C content, and evidence of usage by prehistoric populations. Table 7 shows the species chosen, and the range and median values for their iron and vitamin C contents. The data are in milligrams per 100-gram edible portion. Plants were classified as "high" in iron and/or vitamin C if their median values exceeded the corn median values. The vast majority of "high" iron plants contain well over two times as much iron as corn. All but one of the "high" vitamin C plants contain at least three time as much vitamin C as corn.

The wild plants were grouped into the three categories shown in Table 7. The values for beans and corn are given at the bottom, corn for reference and beans for a separate analysis. Our assumption in the following analysis is that the characteristic of being high in iron and/or vitamin C content relative to corn is a primary factor in a plant's selection.

In the following sections we evaluate two hypotheses concerning the use of plants to offset low meat availability. The first deals specifically with beans. As Table 7 shows, beans are an excellent source of iron. Beans also provide amino acids complementary to those found in corn. We expect that bean ubiquity should be higher at sites with low artiodactyl indices because beans' amino acids complement the amino acid content of corn, thereby improving the quality of plant protein in the diet. Table 7 also shows that certain wild plants are good sources of iron and/or vitamin C. Thus, our second expectation is that ubiquities of plants rich in iron and vitamin C will vary inversely with artiodactyl indices.

TABLE 7 Iron and vitamin C contents of Southwestern wild and cultivated plants.[1]

Category Plant	Range Iron	Range Vitamin C	Median Iron	Median Vitamin C
High in Iron and Vitamin C				
Amaranthus	1.6–22.9	3–120	5.4	72.0
Chenopodium	0.7–37.4	1–109	6.4	7.0
Artemesia	1.5–32.3	36–80	17.3	58.0
Portulaca	1.0–3.6	12–58	3.5	25.0
Solanum	0.3–15.5	4–79	9.9	21.0
Sphaeral.	12.7	35	12.7	35.0
Lepidium	1.3–28.6	19–87	15.0	79.5
High in Iron				
Helianthus	0.4–7.7		7.1	
Pinus	3.1–6.4		5.2	
Prunus	0.3–5.2		4.6	
High in Vitamin C				
Opuntia		14–26		21.0
Yucca		26–393		210.0
Referents				
Beans	1.3–9.2	1–110	3.9	7.0
Corn	0.7–6.2	0–12	2.4	6.2

[1] Values are in mg./100-gram edible portion.

[2] Duke and Atchley (1986).

[3] Cummings (1994).

[4] Instituto de Nutricion (1961).

BEANS

Table 8 presents the data available on bean ubiquities. At several sites no macrobotanical evidence of beans was found. Ubiquity data have not been published for domesticated plants from the village sites at Chaco Canyon.

With beans uniformly low in ubiquity across the Southwest, the data indicate no meaningful trend in bean ubiquities with respect to the artiodactyl indices. There does appear to be much greater recovery of beans at eastern pueblo sites than at western ones, however. It is unclear whether this greater recovery reflects greater consumption or better conditions for preservation at eastern pueblo sites. Beans tend to decay rapidly once discarded, and the generally low bean ubiquities are probably not representative of the contribution of beans to prehistoric Southwestern diets. Variability across the Southwest in the consumption of beans is probably masked due to postdepositional decay. The data at present, however, do not support the hypothesis that there should be greater emphasis on bean cultivation in areas with less abundant large game.

WILD PLANTS

Table 9 provides ubiquity data for the three wild plant categories at the ten sites. In cases where Chenopodium and Amaranthus were reported together, the ubiquity figure was assigned to each species. Figures 4 and 5 show the plots of the ubiquities for each category. In both the table and the figures, the sites are arranged in descending order of artiodactyl index. The line connecting the points was drawn as a visual device to enhance pattern recognition only; it is not meant to suggest any interpolation of ubiquities between sites. The values plotted in Figure 4 are the summation of ubiquity values for all species in each category found at a site. Figure 5 plots the average of the ubiquities of the species in each category found at that site.

The ubiquity pattern of plants high in both iron and vitamin C conforms well with the expected pattern. In contrast, ubiquities of plants that are high in just one of these nutrients do not pattern clearly, but instead seem to vary independently from the artiodactyl index. This is not surprising given that iron from plants is much less bioavailable in the absence of vitamin C. Prehistoric Southwestern farmers appear to have gathered wild plants that provided both iron and vitamin C in the same "package."

To assess the degree of correlation between the artiodactyl index and both the average and the sum of the ubiquities of plants high in both iron and vitamin C, we calculated Spearman's rank order correlation coefficient, rho. Rho is a non-parametric measure of correlation similar to Pearson's r. The values for rho range from 1, a perfect correlation, through 0, no correlation, to -1, a perfect inverse correlation. For the average ubiquity rho was -0.915 and for the sum of ubiquities rho was -0.806. These values indicate a strong inverse relationship between the ubiquities of plant species high in iron and vitamin C and the artiodactyl index.

TABLE 8 Bean ubiquities at aggregated sites.[1]

Site	Bean Ubiquity[2]
Mimbres Salado	2.3
Rowe	0.0
Pueblo Colorado	12.0
Guadalupe	0.0
Gran Quivira	10.0
Mimbres Classic	0.0
Galaz	0.0
Arroyo Hondo	8.6
Pueblo Alto	2.0

[1] Sites are listed in descending order of artiodactyl index value.

[2] Bean ubiquity = No. of samples containing beans/total no. samples.

TABLE 9 Ubiquities[1] of wild plants at aggregated sites.[2]

Site	N	I+C avg.	sum	n	I avg.	sum	n	C avg.	sum
M. Salado	4	24.4	97.6	1	2.3	2.3	0	0.0	0.0
Rowe Pueblo	3	15.6	46.7	0	0.0	0.0	0	0.0	0.0
P. Colorado	4	20.5	82.0	2	15.5	31.0	1	35.0	35.0
Guadalupe	2	30.6	61.1	2	25.0	50.0	2	13.9	27.8
Gran Quivira	3	24.7	74.0	1	12.0	12.0	2	7.0	14.0
M. Classic	4	30.7	122.9	1	14.3	14.3	2	5.8	11.5
Galaz Ruin	4	39.0	156.0	3	8.3	25.0	1	13.0	13.0
Arroyo Hondo	3	36.2	108.6	3	2.5	7.5	2	1.8	3.5
Pueblo Alto	5	37.2	186.0	2	22.5	45.0	2	7.5	15.0
Chaco Villages	4	51.6	206.2	2	11.2	22.3	2	3.6	7.2

[1] The average ubiquity for each wild plant category found at a site was calculated by adding the ubiquities for each species within a category and dividing that figure by the number of species in that category at the site. The sum of ubiquities for each wild plant category found at a site was calculated by adding the ubiquity values for each species within a category.

[2] I + C = High iron and vitamin C plants.
I = High iron plants.
C = High vitamin C plants.

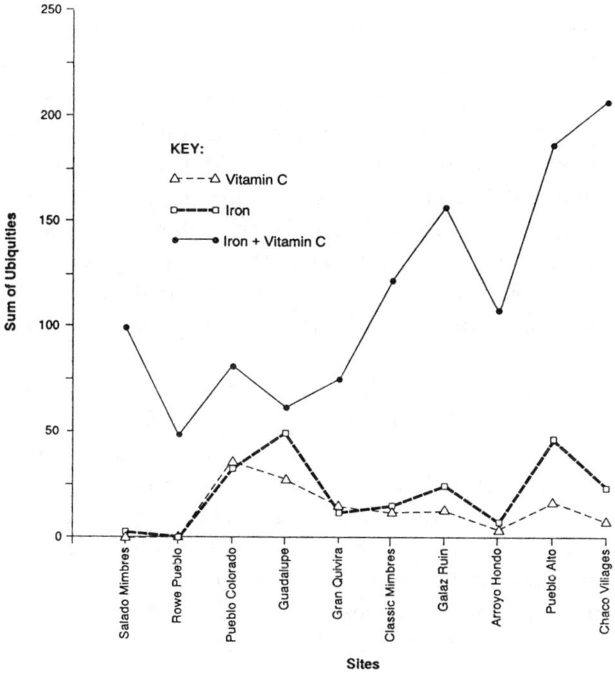

FIGURE 4 Summation of ubiquity values for plant categories.

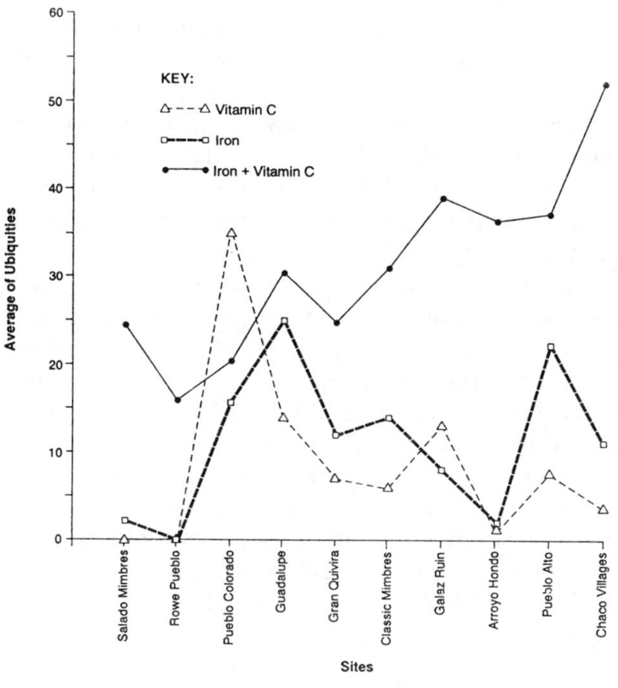

FIGURE 5 Average of ubiquity values for plant categories.

CONCLUSIONS

In this chapter we have compiled a preliminary body of data concerning a variety of strategies that prehistoric aggregated farmers in the Southwest might have chosen when faced with relatively low access to large game. While trade in meat remains a possibility, it is not well documented beyond the eastern border of the Pueblo world. In part this is because such trade is difficult to identify when the species traded are the same as those hunted locally, and if meat is traded largely without bone. Trace element and isotopic analysis of faunal bone from archaeological sites will be necessary to document movement of animal bone across the Southwestern landscape.

Our initial findings are that turkeys were definitely an important source of protein for many late prehistoric Rio Grande and Colorado Plateau farmers. The exceptions were the Salinas Pueblos, who may have obtained adequate supplementary meat through trade. The Sonoran Desert farmers did not raise turkeys, and very small game is likely to have been a more important source of protein for them.

We also considered plants that could have been used to supplement a meat-poor diet. With respect to beans, there is no evidence for a greater emphasis on bean production at sites with low artiodactyl indices, despite the fact that beans provide amino acids that are missing in corn and could make up for the deficit in iron that a lack of meat would have created.

There is, however, compelling evidence that wild plants higher than corn in iron and vitamin C content were more extensively harvested at sites with lower artiodactyl indices. It would be unwise, however, to place too much emphasis on the iron and vitamin C content of these plants without further information. It will be important, for example, to determine whether ubiquities of these wild plants correlate directly with their densities in the wild. Moreover, plants provide other important nutrients, such as calories, which may also have been dietarily significant, particularly at times when corn was in short supply. Thus, the use of wild plants is likely to have satisfied multiple nutritional needs.

To conclude, variability in access to large game appears to have been significant across the Southwest. While this variability may not have had the impact on prehistoric populations that success or failure in crop production had, nonetheless several strategies appear to have been developed to cope with relatively limited access to meat. These strategies have important implications for diet, health, and subsistence organization in the Southwest, which will require interdisciplinary research efforts to document and explain. For example, an analysis of the health of individuals from each of these sites would allow an evaluation of our conclusions based on dietary data alone. We look forward to a time when in-depth bioarchaeological analyses (e.g., Martin et al. 1992) and synthetic studies of prehistoric dietary data (e.g., Wetterstrom 1986) can be integrated into a more complete and accurate assessment of diet and health in the prehistoric Southwest.

ACKNOWLEDGMENTS

Numerous people have assisted us in this endeavor. In particular we would like to thank Dr. Carol N. Angstadt for providing us with a multitude of nutritional references, Linda Cordell for providing data from the Rowe Pueblo excavations, Paul Minnis for the loan of his reference on food nutrients, and Margaret Nelson for providing faunal data from several Mimbres sites. Our spouses, Stefani A. Angstadt-Leto and William C. Edwards, were of great help in completing the manuscript.

This research also benefitted tremendously from the week-long SFI conference, at which it was first presented. The intensity and quality of the interaction and information at this working session were truly remarkable, and we thank the organizers for inviting our participation.

REFERENCES

Akins, Nancy

1985 Prehistoric Faunal Utilization in Chaco Canyon: Basketmaker III through Pueblo III. In *Environment and Subsistence of Chaco Canyon*, edited by F. J. Mathien, pp. 305–445. Publications in Archaeology 18E. USDI, National Park Service.

1987 Faunal Remains from Pueblo Alto. In *Investigations at the Pueblo Alto Complex, Chaco Canyon*, Volume III, pt. 2, edited by F. J. Mathien and T. C. Windes, pp. 445–649. Publications in Archaeology 18F. USDI, National Park Service.

Anyon, Roger, and Steven LeBlanc

1984 Faunal Remains. In *The Galaz Ruin: A Prehistoric Mimbres Village in Southwestern New Mexico*, edited by R. Anyon and S. LeBlanc, pp. 215–223. Maxwell Museum of Anthropology and the University of New Mexico Press, Albuquerque.

Banerji, L., S. K. Sood, and V. Ramalingaswami

1968 Geographic Pathology of Iron Deficiency with Special Reference to India. *American Journal of Clinical Nutrition* 21:1139–1148.

Bayham, Frank

1982 *A Diachronic Analysis of Prehistoric Animal Procurement at Ventana Cave.* Ph.D. dissertation, Arizona State University, Tempe. University Microfilms, Ann Arbor.

Bayham, Frank, and Pamela Hatch

1985 Archaeofaunal Remains from the New River Area. In *Hohokam Settlement and Economic Systems in the Central New River Drainage, Arizona*, edited by D. Doyel and M. Elson, pp. 405–433. Publications in Archaeology No. 4. Soil Systems, Phoenix.

Baynes, R. D., and T. H. Bothwell
1990 Iron Deficiency. *Annual Review Nutrition* 10:133–148.
Cummings, Linda Scott
1994 Anasazi Diet: Variety in the Hoy House and Lion House Coprolite Record and Nutritional Analysis. In *Paleonutrition: The Diet and Health of Prehistoric Americans*, edited by K. D. Sobolik. Occasional Paper 22. Center for Archaeological Investigations, Southern Illinois University, Carbondale.
Dagnelie, P. C., W. A. Van Staveren, F. J. V. R. A. Vergote, J. Burema, and M. A. Van't Hof
1989 Nutritional Status of Infants Aged 4 to 18 Months on Macrobiotic Diets and Matched Omnivorous Control Infants: A Population-Based Mixed Longitudinal Study. 2. Growth and Psychomotor Development. *European Journal of Clinical Nutrition* 43:325–388.
Dagnelie, P. C., W. A. Van Staveren, F. J. V. R. A. Vergote, P. G. Dingjan, H. Van Den Berg, and J. G. A. J. Hautvast
1989 Increased Risk of Vitamin B12 and Iron Deficiency in Infants on Macrobiotic Diets. *American Journal of Clinical Nutrition* 50:818–824.
Dagnelie, P. C., F. J. V. R. A. Vergote, W. A. Van Staveren, H. Van Den Berg, and J. G. A. J. Hautvast
1990 High Prevalence of Rickets in Infants on Macrobiotic Diets. *American Journal of Clinical Nutrition* 51:202–208.
Diehl, Michael
1990 Analysis of Charred Plant Remains from Pueblo Colorado (LA476), New Mexico. Report 22. Ethnobotanical Laboratory, University of Oklahoma.
Duke, James A., and Alan A. Atchley
1986 *CRC Handbook of Proximate Analysis Tales of Higher Plants.* Boca Raton: CRC Press.
Dwyer, J. T.
1991 Nutritional Consequences of Vegetarianism. *Annual Review of Nutrition* 11:61–91.
Dwyer, J. T., W. H. Dietz, E. M. Andrews, and R. M. Suskind
1982 Nutritional Status of Vegetarian Children. *American Journal of Clinical Nutrition* 35:204–216.
El-Najjar, M. Y., D. J. Ryan, C. G. Turner, and B. Lozoff
1976 The Etiology of Porotic Hyperostosis Among the Prehistoric and Historic Anasazi Indians of the Southwestern United States. *American Journal of Physical Anthropology* 44:477–488.
Ezzo, Joseph
1992 Dietary Change and Variability at Grasshopper Pueblo, Arizona. *Journal of Anthropological Archaeology* 11:219–289.
FAO/WHO
1973 *Energy and Protein Requirements: A Report of a Joint FAO/WHO Ad Hoc Expert Committee.* Food and Agriculture Organization Nutrition Meeting

Report Series 520. World Health Organization Technical Report Series 522.

Goodman, Alan

1994 Cartesian Reductionism and Vulgar Adaptationism: Issues in the Interpretation of Nutritional Status in Prehistory. In *Paleonutrition: The Diet and Health of Prehistoric Americans*, edited by K. D. Sobolik. Occasional Paper 22. Center for Archaeological Investigations, Southern Illinois University, Carbondale.

Hallberg, Leif

1981 Bioavailability of Dietary Iron in Man. *Annual Review Nutrition* 1:123–147.

Instituto de Nutricion de Centro America y Panama.

1961 *Tabla de Composicion de Alimentos para Uso en America Latina.* Guatemala: Ciudad Guatemala.

James, Steven

1987 Hohokam Faunal Resources in the Waddell Project Area. In *Waddell Report*, edited by Margery Green. Archaeological Consulting Services, Tempe, AZ.

Keene, Arthur

1981 *Prehistoric Foraging in a Temperate Forest: A Linear Programming Model.* Academic Press, New York.

Kuhn, I. N., M. Layrisse, M. Roche, C. Martinez, and R. B. Walker

1968 Observations on the Mechanism of Iron Absorption. *American Journal of Clinical Nutrition* 21:1184–1188.

Lang, Richard W., and Arthur H. Harris

1984 *The Faunal Remains from Arroyo Hondo Pueblo.* School of American Research Press, Santa Fe.

Layrisse, M., C. Martinez-Torres, and M. Roche

1968 Effect of Interaction of Various Foods on Iron Absorption. *American Journal of Clinical Nutrition* 21:1175–1183.

McKusick, Charmion

1981 The Faunal Remains of Las Humanas. In *Contributions to Gran Quivira Archaeology*, edited by A. Hayes, pp. 39–65. Publications in Archaeology 17. USDI, National Park Service, Santa Fe.

Martin, D. L., A. H. Goodman, G. J. Armelagos, and A. L. Magennis

1991 *Black Mesa Anasazi Health: Reconstructing Life from Patterns of Health and Disease.* Occasional Paper 14. Center for Archaeological Investigations, Southern Illinois University, Carbondale.

Matson, R.G., and B. Chisholm

1991 Basketmaker II Subsistence: Carbon Isotopes and Other Dietary Indicators from Cedar Mesa, Utah. *American Antiquity* 56:444–459.

Mick O'Hara, Linda
1987 *Identification and Analysis of the Faunal Remains from the 1984 Excavations at Rowe Pueblo.* Rowe Ruin Project: Preliminary Report. Ms. in possession of the author.

Minnis, P. E.
1984 Macroplant Remains. In *The Galaz Ruin: A Prehistoric Mimbres Village in Southwestern New Mexico*, edited by R. Anyon and S. LeBlanc, pp. 193–200. Maxwell Museum of Anthropology and the University of New Mexico Press, Albuquerque.
1985 *Social Adaptations to Food Stress: A Prehistoric Southwestern Example.* Chicago: University of Chicago Press.
1986 Macroplant Remains. In *Short-term Sedentism in the American Southwest: The Mimbres Valley Salado*, by B. Nelson and S. LeBlanc, pp. 205–218. Maxwell Museum of Anthropology and the University of New Mexico Press, Albuquerque.
1989 Prehistoric Diet in the Northern Southwest: Macroplant Remains from Four Corners Feces. *American Antiquity* 54:543–563.

Nelson, B. A., D. L. Martin, A. C. Swedlund, P. R. Fish, and G. J. Armelagos
1992 Studies in Disruption: Demography and Health in the Prehistoric Southwest. Paper presented at the 57th Annual Meeting of the Society for American Archaeology, Pittsburgh.

Palkovich, A.
1980 *The Arroyo Hondo Skeletal and Mortuary Remains.* School of American Research Press, Santa Fe.

Peterson, Jane
1989 Changing Turkey Utilization at Gran Quivira Pueblo. Ms. in possession of author.

Pippin, Lonnie C.
1987 *Prehistory and Paleoecology of Guadalupe Ruin, New Mexico.* Anthropological Papers 112. University of Utah, Salt Lake City.

Raymer, Leslie, and Paul Minnis
1992 Ethnobotanical Relationships at Gran Quivira. In *Subsistence and Exchange at Gran Quivira Pueblo*, by Katherine A. Spielmann. Report submitted to the USDI, National Park Service, Southwest Regional Office, Santa Fe.

Speth, John D., and Susan L. Scott
1989 Horticulture and Large Mammal Hunting: The Role of Resource Depletion and the Constraints of Time and Labor. In *Foragers as Farmers: The Implications of Sedentism*, edited by Susan Kent, pp. 71–79. Cambridge, MA: Cambridge University Press.

Spielmann, Katherine A.
1983 Late Prehistoric Exchange Between the Southwest and the Southern Plains. *Plains Anthropologist* 28:257–272.

1988 Changing Faunal Procurement Strategies at Gran Quivira Pueblo, New Mexico. Paper presented at the 53rd Annual Meeting of the Society for American Archaeology, Phoenix.

1991a Protein Acquisition in the American Southwest. Paper presented at the 56th Annual Meeting of the Society for American Archaeology, New Orleans.

1991b *Interdependence in the Prehistoric Southwest.* New York: Garland Publishing.

1991c *Farmers, Hunters, and Colonists: Interaction Beetween the Southwest and Southern Plains.* Tucson, AZ: University of Arizona Press.

Spielmann, K. A., and M. J. Schoeninger

1992 Multidisciplinary Studies of Trade in Meat at Gran Quivira Pueblo, New Mexico. Paper presented at the conference, Paleonutrition: The Diet and Health of Prehistoric Americans, Southern Illinois University, Carbondale.

Spielmann, K. A., M. J. Schoeninger, and K. Moore

1990 Plains-Pueblo Interdependence and Human Diet at Pecos Pueblo, New Mexico. *American Antiquity* 55:745–765.

Stodder, Ann L. W.

1990 *Paleoepidemiology of Eastern and Western Pueblo Communities in Protohistoric New Mexico.* Ph.D. dissertation, University of Colorado, Boulder.

Stuart-Macadam, P.

1987 Porotic Hyperostosis: New Evidence to Support the Anemia Theory. *American Journal Physical Anthroipology* 74:521–526.

Szuter, Christine

1989 *Hunting by Prehistoric Horticulturalists in the American Southwest.* Ph.D. dissertation, University of Arizona, Tucson. University Microfilms International, Ann Arbor.

1992 Agriculture, Hunting, and Gathering: Human Responses to Anthropogenic Environmental Change. Paper presented at the Workshop on Resource Stress, Economic Uncertainty, and Human Response in the Prehistoric Southwest, Santa Fe Institute, February 25–29.

Szuter, C., and Frank Bayham

1989 Sedentism and Prehistoric Animal Procurement Among Desert Horticulturalists of the North American Southwest. In *Foragers as Farmers: The Implications of Sedentism*, edited by Susan Kent, pp. 80–95. Cambridge, MA: Cambridge University Press.

Thiel, J. Homer

1994 The Effects of Sedentism on Faunal Resources at Pueblo Colorado. Ms. in possession of the author.

Toll, Mollie S.

1981 A Preliminary Investigation of Botanical Remnants of Subsistence at Rowe Pueblo. Unpublished Ms. in possession of Linda Cordell and the authors.

Toll, Mollie S.
1985 An Overview of Chaco Canyon Macrobotanical Materials and Analyses to Date. In *Environment and Subsistence of Chaco Canyon*, edited by F. Mathien, pp. 247–277. Publications in Archaeology 18E. USDI, National Park Service, Santa Fe.
1987 Plant Utilization at Pueblo Alto: Flotation and Macrobotanical Analyses. In *Investigations at the Pueblo Alto Complex, Chaco Canyon*, Volume III, pt. 2, edited by F. J. Mathien and T. C. Windes, pp. 691–784. Publications in Archaeology 18F. USDI, National Park Service, Santa Fe.

Wetterstrom, Wilma
1986 *Food, Diet, and Population at Prehistoric Arroyo Hondo Pueblo, New Mexico.* Arroyo Hondo Archaeological Series, Vol. 6. School of American Research Press, Santa Fe.

Windes, Thomas C.
1987 The Use of Turkeys at Pueblo Alto Based on the Eggshell and Faunal Remains. In *Investigations at the Pueblo Alto Complex, Chaco Canyon*, Volume III, pt. 2, edited by F. J. Mathien and T. C. Windes, pp. 679–690. Publications in Archaeology 18F. USDI, National Park Service.

Margaret C. Nelson
SUNY Buffalo, Department of Anthropology, Buffalo, NY 14261

Technological Strategies Responsive to Subsistence Stress

Subsistence stress is a common explanation for changes in economic and social patterns inferred from the archaeological record of the prehistoric Southwest. While some attention has been paid to technological strategies for intensifying cultivation, little effort has been made to go beyond this limited aspect of technology, and explore how a range of technological behaviors may have been responsive to subsistence stress. In this chapter, I discuss how strategies of technological behavior may address the risks of shortfall in provisioning people with food resources. I examine the technology of daily life rather than major technological systems, such as irrigation or transportation systems, although the strategies that I discuss should be relevant to analysis of these. This chapter explores new ways of examining technology; my discussion combines ecological notions about economic responses to provisioning risks and resource stress with theories about the design of technology.

I use the term risk along with stress because stress is the realization of potential risks of a pattern of land use. The literature of evolutionary ecology contains abundant information about how animals respond to risk and many models of optimal responses, some of which have been applied to understanding human behavior (e.g., Cashdan 1990). In human societies, technological knowledge as well as other cultural domains of information add dimensions to the potential for humans to respond to risk and stress. I focus on technology because it is the facilitator of many social and economic decisions about how to adapt to and manipulate the environment. These decisions are implemented through the construction and use of houses, hearths, tools, weapons, containers, among many classes of material culture.

This chapter has three parts. The first establishes the context of the study with a consideration of some of the risks or potential points of subsistence stress of prehistoric hunting, gathering, and cultivation in the North American Southwest. This section also includes a review of ecological views on potential responses to these kinds of stresses. In the second section, I examine possible technological behaviors that facilitate the responses to risk identified in the first section. Every effort is made to discuss the material implications of these technological behaviors. In discussing tools and weaponry, implications for the form of the stone elements are emphasized. The final section is an examination of changes in the manufacture and use of stone-tipped weapon technology during late, ceramic-period occupations in the Southwest as evidence of technological strategies for responding to subsistence risks. Preceding these three sections is a brief discussion of terminology.

DEFINITIONS

Risk is defined in many different ways (see Cashdan [1990] for examples and much discussion), but I use Winterhalder's discussion of subsistence risk as a definition: "...the possibility of harmful shortfalls...induced by unpredictable or stochastic factors in [the] environment" (1990:67). As pointed out by Hegmon (1989:90) and by Bamforth and Bleed (1991), risk has two aspects: the probability of failure and the cost of the loss. Torrence (1989:59) refers to the latter as the severity of loss and notes that "...the level of investment into technology will be determined by the severity of the consequences of losing the resource," among other variables. In this volume, our discussions focus on loss that is costly enough to result in subsistence stress. The role that technology plays in managing risk and resolving stress is in reducing the probability of failure (Bamforth and Bleed 1991). "As the consequences of the loss of a resource increase, one can expect that more effort will be put to devising technology for exploiting it" (Torrence 1989:59), that is for reducing the probability of failure.

Technological organization, a phrase used primarily among lithic technologists, refers to the relationship among strategies for manufacturing, manipulating (using, reusing, reshaping), and abandoning (losing, discarding) material items (Nelson 1991:58 from Koldehoff [1987:154], Kelly [1988:717], and Binford [1977]). It includes the actors' knowledge about, as well as behavior within, these domains (Schiffer and Skibo 1987). The knowledge base of any technology is considerably broader than the behavior exemplified at any one point in time. Behavior changes through technological innovation (change in the knowledge base), or by modifications of economic and social or other cultural strategies that influence the implementation of particular technological strategies from the pool of available knowledge. The stimuli for economic and social change are many, but in this volume we have chosen to examine stress in provisioning people with food as one important stimulus.

From a systemic perspective, technology and these other social, economic, political, and ideological domains are interrelated components of cultural systems. While archaeologists have given their attention to major innovations in technology (e.g., domestication and irrigation), both selection and innovation in the more mundane technological strategies employed in manufacturing, using and abandoning items (ceramics, stone, and bone tools), or features (hearths and houses) have received relatively little attention. Such mundane technological strategies become more interesting when we realize the ways that their accomplishment is determined by other cultural domains. For example, horticulturalists who are seasonally mobile may design, produce, use, and abandon tools and houses in ways that are quite different from those of horticulturalists who remain residentially stable throughout the year. Similarly, groups with restricted social networks may design, produce, use, and discard trade items in ways different from groups with extensive social networks. Technological strategies intervene between other cultural strategies and the surrounding material conditions.

We cannot assume that people always act(ed) efficiently, optimally, or even in their own interest. When resources are relatively abundant, it is difficult to model how people may have behaved except at the most general level of provisioning themselves sufficiently. It is difficult to model technological strategies because the costs of technological failure are low under conditions of resource abundance or lack of food stress (Bamforth and Bleed 1991). However, at times when resources are limited, not only is the cost of failure increased, but time spent making tools, weapons, and facilities may conflict with important time spent acquiring or producing food (Bamforth and Bleed 1991). Under these conditions, the technological options are clearer. "Technology... represents a solution to the problem of managing risk" (Torrence 1989:57), of reducing the probability of technological and, therefore, subsistence failure.

SOME ECONOMIC AND SOCIAL RESPONSES TO SUBSISTENCE RISK

In order to discuss technological changes that are responsive to subsistence stress it is necessary to identify the kinds of stress and possible social and economic responses that occurred in the prehistoric North American Southwest. These include the possible stressors affecting hunting, gathering, and cultivation at various levels of investment.

The North American Southwest is a seasonal environment, with the most extreme variability evident in montane areas because of the varied topography. It is also an environment of relatively low productivity, primarily due to low precipitation levels throughout much of the region. Precipitation and, therefore, resource productivity, varies among mountain, plateau, and desert settings (Cordell 1984).

For hunter-gatherers, environmental seasonality requires a shifting annual cycle of resource selection (Kelly 1983; Low 1990). It may also imply that high elevations cannot be occupied during winter, without sufficient storage, because of the paucity of resources. The low productivity of most regions in the Southwest indicates that for hunter-gatherers, resource stress can be brought on by population increases or by periods of especially low productivity due to climatic change or fluctuation (drought, freezing).

For cultivators who hunt and gather, seasonality implies a limited time frame for cultivation, requiring food storage and dependence on noncultivated resources for some portion of the annual cycle. Under the general conditions of relatively low average precipitation and natural productivity, stress can occur with population increase, decline in soil productivity, or climatic change (short or long term), among many other factors. The occurrence of such conditions and their effects have been discussed in great detail in the ethnographic and archaeological literature of the Southwest (e.g., Graves et al. 1982; Hegmon 1989; Minnis 1985a; Rautman 1993).

In the literature of evolutionary ecology and archaeology dealing with stress and risk of shortfall, a variety of responses are considered. These include *specialization*, *diversification*, and *risk pooling*, among others. *Resource specialization* is a focus on one or a few resources at the expense of others. This strategy is implemented to increase the upper range of potential returns from that resource and, therefore, raise the mean expected returns to bring them up to the threshold of needs. Winterhalder (1990:76) argues that restricted diet breadth (specialization) should occur under conditions of higher energy requirements than can be expected from the environment on average. With this strategy, the actors may become variance-prone (Hegmon 1989:93) as a consequence of increasing the likelihood of being above the threshold of food needs some of the time. In the Southwest, maize is one of the few, or perhaps only, resources that can become the primary focus of subsistence activities. One or a few naturally occurring plants and animals may not be productive enough to become the focus of subsistence, and archaeologists have argued that prehistoric hunter-gatherers moved seasonally between resource zones (Wills 1988). In a cross-cultural study, Low (1990:247) found that storage is most common in predictable environments. Maize may be the most predictable resource available in the Southwest, and Wills (1988) documents an increase in use of storage facilities with the early adoption of maize cultivation.

Diversification can occur in two ways. Groups may stay focused on one or a few resources and diversify their procurement techniques, such as planting in multiple environments (Winterhalder 1990; Hegmon [1989] summarizes literature on the Hopi: Minnis [1985a] describes this for the prehistoric Mimbres; Fish, Fish, and Madsen [1985] describe prehistoric Hohokam in the Tucson Basin; Graves et al. [1982] note this practice in the Point of Pines area), distributing planting times (Hegmon [1989] for the Hopi), planting different varieties of a kind of cultigen

(Baksh and Johnson 1990; Hegmon 1989), employing direct and indirect techniques for capturing game or fishing (Baksh and Johnson 1990), or trading for a target food (Graves et al. 1982; Low 1990; Spielmann 1983). These strategies are employed in conjunction with specialization on one or a few kinds of resources. Alternatively, groups may diversify by broadening the kinds of resources that are exploited (Baksh and Johnson 1990; Reid 1978; Winterhalder 1990). This reduces the effect of temporal variation in productivity of any one resource. However, expanding diet breadth may reduce procurement efficiency and, therefore, reduce the mean intake of individuals (Hames 1990:95; Winterhalder 1990). If an increase in diet breadth can be accomplished without decreasing the time spent on previously procured resources, and therefore the yields from them, then mean intake may be increased even though efficiency is reduced.

Specialization and diversification are economic decisions intended to ameliorate stress or reduce the risk and cost of shortfall. Implementation of any one strategy depends on a variety of ecological and cultural conditions that I do not explore here. Rather, I examine the technological strategies that would be evident given the implementation of one or the other of these economic strategies.

There are also social strategies for dealing with stress or risk of shortfall (Braun and Plog 1982; Graves at al. 1982; Hegmon 1989; Rautman 1993; Wiessner 1982, 1983). One of these is *pooling resources or "pooling risk" within the resident group* (Baksh and Johnson 1990; Cashdan 1980; Hegmon 1989; Winterhalder 1990). Those who study foraging societies have documented extensive, generalized intra-group sharing and reciprocity (e.g., Cashdan 1980, 1983; Woodburn 1982), while the intra-group sharing patterns of cultivators and sedentary hunter-gatherers are more limited (Baksh and Johnson 1990; Hegmon 1989:92). According to Winterhalder, pooling is an effective way to circumvent the shortfalls of individual diet selection, but "major gains in risk reduction occur for relatively small group size" (1990:79; see also Hegmon [1989:104–105]).

Within foraging societies, conditions of the environment influence sharing group size. Cashdan (1983) has argued that for hunter-gatherers, the productivity of the natural environment is correlated with social access to the resources of that area. Where productivity is low and resource areas are large, social access is restricted. This is accomplished not through defense of physical territory, but through social limits on who has access to the resource pool. In other words, under conditions of fairly high risk of shortfall, the composition of the sharing group is restricted.

Hames (1990:96–97) argues that cultigens may be shared less often than wild resources, especially game, for at least two reasons. First, the risk in production of cultigens is less than the risk in hunting. Second, preservation is less costly for cultigens than for meat. Thus, storage for individual households is a more viable option for cultivators to hedge against shortfall than it is for hunters. Although Hames' opinion is derived from his study of tropical horticulturalists, his conclusion may apply to the Southwest, as well, if cultigens are a more predictable resource than are game. It has been argued that cultivation developed in the Southwest to enhance the predictability of food resources and reduce subsistence risk (Cordell and Gumerman 1989:8–9; Hunter-Anderson 1986; Wills 1988; Wills and Huckell 1994). While the productivity of cultigens is influenced by the unpredictability of precipitation, so is the productivity of game. The location and potential yield of cultigens is managed by groups, while the location and abundance of game is generally not.

Hegmon (1989) argues that sharing among prestate, nonpeasant cultivators is an important strategy for responding to risk, but that the size of sharing groups is small as among hunter-gatherers. In various simulations, modeling the effects of possible social responses to food shortfalls, she uses a group size of five households (effectively 35 people) and finds that restricted sharing (only of household surplus) reduces the risk of shortfall more substantially than does pooling of all yields, except under the most extreme conditions (see also Hegmon [1995]). Comparing the different effects of time delays for resource returns on sharing patterns between hunting-gathering and cultivating groups, Winterhalder (1990) argues that for resources with considerable delay in returns, such as cultigens, pooling and sharing are problematical. It is too easy, he argues, for some individuals to shirk their responsibilities to the group when there is no regular short-term assessment of who is contributing, as occurs with daily hunting and gathering. Thus, while intra-group sharing occurs within both hunter-gatherer and cultivator groups (at least those in prestate, nonpeasant societies—see Hegmon [1989]; Winterhalder [1990]), sharing groups must be relatively small and sharing is less generalized among cultivators.

Another response to actual or potential shortfall is to *pool and share resources or risk over a broad area.* This may be accomplished by maintaining broad regional contacts and reciprocal access (movement or trade) with areas that are expected to be productive when one's own region is experiencing a resource shortfall (Smith and Boyd 1990; Rautman 1993[1]; Winterhalder 1990:79). Low (1990:247) argues on the basis of a cross-cultural study of environmental risk that the more unpredictable in time are fluctuations in food supply, the more broadly-based regional trade alliances should be. Smith and Boyd (1990) model the costs and benefits of long-distance sharing. They observe that moving resources is considerably more costly than moving people. Their model focuses on the latter, considering the conditions under which allowing access to one's territory is cost effective. They find that three factors influence the decision to allow access: payoffs must be monotonically increasing as a function of food harvest, the cost of defense must be less than the cost of allowing access, and the probability of the neighbor's area failing must not be high (Smith and Boyd 1990:187). Rautman (1993) has emphasized the third criterion, using contemporary climatic data to develop models of appropriate access areas for prehistoric horticulturalists in the Gran Quivira area (see also Rautman 1995).

While social strategies for reducing risk are at least as important as economic responses, in the next section I restrict my consideration of technological strategies to those that facilitate the two economic strategies described above: specialization

[1]See also the chapter by Rautman in this volume

and diversification. Less attention has been paid to the relationship between technological and social strategies, but the work of Wobst (1977), Plog (1978), Hill (1985), and Wiessner (1982, 1983), among others, can be used to begin modeling relationships relevant to subsistence stress and risk.

TECHNOLOGICAL STRATEGIES

To understand how technological strategies reduce the probability of failure under conditions of risk and stress, four design concepts are used which combine considerations of tool and facility production and use: *reliability*, *flexibility*, *versatility*, and *portability* (for a detailed discussion see Bleed 1986; Nelson 1991; Shott 1986; Torrence 1990). *Reliability* directs production and use toward the lowest probability of technological failure, avoiding breakage and assuring fit between task and tool. High investments are made in production and tools are expected to have long use-lives. *Flexibility* emphasizes lowering the probability of resource loss by building in the capability for a tool or facility to change form according to variable needs. *Versatility* serves a purpose similar to that of flexibility, but the design is multifunctional or generalized without changes in form. *Portability* directs production toward designs that result in tools that can be moved easily so they are available when needed. This minimizes the risk of loss from lack of available technology. All of these options contribute to effective use of resources, but in different ways. Similarly, all design variables are beneficial, but all have costs. The optimal tool form combines these design options in ways that suit economic and social strategies. It is a juggling game of weighing advantages and disadvantages. At the least, all tools and facilities must be designed to perform the work to which they are applied. This can be accomplished in many ways depending on a variety of other conditions for tool design including (but not exclusively) mobility strategies, availability of resources, physical context of work, social context of work, in addition to risk. While I discuss responses to risk and stress, these should not be considered the only conditioners of technological strategies.

In this section, I discuss the kinds of technological strategies that would facilitate economic strategies of resource specialization and diversification. Hunting, gathering, and cultivating practices are considered. Cultural conditions are as important as ecological, therefore I target my discussion only to one phase in the prehistory of the Southwest. I address the phase of intensive cultivation, which occurred after A.D. 900, when cultigens formed a substantial portion of the diet (Matson and Chisholm 1986; Minnis 1989). Expectations for hunting and gathering in this context may be different from those for people who hunt and gather exclusively, or who cultivate only casually because of differences in mobility and demography.

SPECIALIZATION

Specialization involves a focus on one or a limited number of resources. Acquiring as much as possible of these few resources is the primary objective. In the Southwest, specialization on cultigens is perhaps most likely, but intensive cultivators may also have narrowed their game selection or wild plant acquisition as a risk minimizing strategy. All of these are explored.

For a specialized resource focus, in general, facilities that improve resource yield or the longevity of resource storage are important (Hegmon 1989:90; Torrence 1989:59; Wiessner 1982:172–173). Tool use-efficiency in capturing and processing is critical for this subsistence strategy because of the emphasis on acquiring as much as possible of a limited range of resources. Tool reliability is important, as well, in order to maximize resource acquisition by minimizing the possibility of tool and weapon failure. These technological strategies would be valuable regardless of the economic strategy, but for resource specialization they are critical.

To improve resource yields in the semiarid environment of the Southwest, access to water is a critical variable. Any specialized subsistence strategy should involve efforts to improve access to water for humans and for the resources targeted, thus reducing risk of subsistence failure. Assess to water can be improved by digging out springs or damming creeks to create small pools, and by channeling water. Pooling and damming should have occurred only on a small scale, as evaporation is a problem. Seeing these technological strategies archaeologically is difficult but not impossible.

Dependable storage is necessary to any specialized resource focus in a seasonal environment such as the Southwest. Watson (1991:191) states that storage is evident in the Southwest as well as other regions prior to extensive or intensive investment in agriculture. Excavation has yielded limited evidence of pit storage (Wills 1988), but no specialized structures or vessels used for storage prior to the adoption of cultigens in the the Southwest. However, Young (1992), in a cross-cultural study of ethnographic documentation on Southwestern native groups, finds that pit storage was common among the most mobile groups as was caching in rockshelters. These strategies allow stores to be concealed when occupants are away from a site. Pottery was used along with skin bags to store food in shelters.

The tools and weapons used by resource specialists should be use-efficient and reliable. "The most [use-]efficient tools are those designed to undertake only a very limited range of functions" (Torrence 1989:61, word added). Thus specialized forms are expected. Such specialization may result in a diverse tool assemblage, which is not easily transportable. With a focus on one or a few resource areas, however, transportability is not a critical variable. Reliability assures that tools and weapons work when needed (Bleed 1986). It is accomplished by investing in the production of items, so that they do not break or fail to work. Design considerations may include selection of special materials, overdesigned parts, and multiple backup elements (Bleed 1986). Reliability is most crucial for hunting weaponry because failure of a tool to work immediately can result in loss of the game (Bamforth and Bleed

1991; Torrence 1989). Tool designs that emphasize specialization and reliability may take longer to manufacture than do those that are generalized because of the emphasis on producing specific forms with durable qualities. Reuse and repair of these specialized, reliable tool forms may be expected in order to compensate for initial investments in manufacture (Binford 1979; Nelson 1991). With such attention to reliability, use-efficiency, and investment of production time, tool and weapon forms may be fairly standardized.

These general statements about technological strategies for increasing productivity, improving storage, and assuring capture and processing of resources are discussed for specialized emphases on cultivation, hunting, and gathering by prehistoric horticulturalists.

CULTIVATION. Water control features greatly enhance the yield of cultigens in the Southwest. Field modification features to compensate for cold air drainage or potential frost also enhance yields. For the late prehistoric phases in the Southwest a variety of field modification and water control facilities suited to many different potential field areas have been documented (e.g., Cordell 1984; Fish, Fish, and Madsen 1985; Herrington 1979; Vivian 1974). For example, in the Mimbres and Point of Pines areas at the height of population aggregation, a wide range of damming and channeling facilities were developed (Graves, et al. 1982; Herrington 1979). Modification of springs has received little attention in the Southwest, but Meltzer (1991) has identified it archaeologically at Mustang Springs on the southern High Plains, indicating that it can be detected. Water channeling and pooling would improve the yields of wild plant and animal resources as well as those of cultigens.

Storage is a necessary technology for specialization on cultigens in the Southwest. Young (1992) notes that increased sedentism, which occurs with development of cultivation, is correlated with more storage, greater variety in storage facilities, and more conspicuous facilities. In the Southwest these might include special structures, rooms, vessels, racks, and pits. Storage rooms are a common puebloan occurrence associated with increased investment in agriculture. Greater attention could be paid to identifying storage vessels and racks.

Ceramic vessels make excellent storage containers because they can be sealed against small animals. Where ceramic vessels were used for storage of cultivated grain, permeability may have been important to protect against spoilage. Vessels can be made permeable through appropriate tempering material, low firing temperature, and lack of surface treatment such as burnishing or smudging (Skibo and Schiffer, personal communication, 1992). It is possible that specialized storage vessels were never produced in the North American Southwest. Among the Maya in San Mateo Ixtatan in the Guatamala Highlands, who invest heavily in maize agriculture, food is stored in vessels that have cracked in manufacture, or use, or are newly manufactured cooking jars, as yet unused for that purpose (personal observation, 1980). However, the context of resource specialization is one in which manufacture of specialized storage vessels might have occurred.

With increasingly specialized economies relying on cultivation, the design of tools associated with cultivation should change toward increased reliability and use-efficiency. Cultivation tools are difficult to see archaeologically because many were wood. At least two kinds of stone tools, however, can be considered: stone axes and grinding stones.

Where felling trees for field clearing and construction of storage facilities is important to increasing emphasis on cultivation, axes should be made to be more reliable and efficient. Axes, as with projectile points, can be made more reliable and efficient through selection of durable material, large size, haft strength, and careful grinding (see Christenson [1987] for projectile points). Dense metamorphic rock is excellent for withstanding the potential shock of impact imposed on an axe. However, stone that can effectively withstand impact is difficult to shape; thus, production time is greater when more durable materials are used to increase the reliability of an axe. Within limits, larger axes provide greater weight behind each blow, decreasing the time needed to fell trees. Larger axes can also be repaired and resharpened more times, improving the use-life and decreasing the number of axes that need to be made. In addition, thickness enhances tool reliability by decreasing the likelihood of breakage during use, although thickness may interfere with the sharpness of a tool (see Nelson [1981] for a discussion of this relationship for stone tools). Haft strength may be improved with attention to the neck of the axe. While I do not have any information specifically on securing axe hafts, Christenson (1987:145–147) has argued that hafting is made secure on projectile points by maximizing contact of the point with the seizing. Thus, on projectile points, the stem and base are made wide. However, projectile points are an extension of their shaft, while axes are hafted perpendicularly to their handle. A deep, full groove may be most amenable to secure hafting for axes. While hafting improves the reliability of an axe, grinding may improve its use-efficiency. Boydston (1989) has described axe finishing by grinding as a time-consuming activity, especially in comparison to chipping tool edges, but he argues that as the uses to which axes are put becomes crucial to subsistence, the investment in grinding is returned in the improved efficiency of the tool. As noted above, the emphasis on production of efficient, reliable designs may result in standardization of axe forms. Such standardization may not be evident in axe length or weight because reuse and reworking are expected (Nelson and Lippmeier 1993). In short, expectations for the design of stone axes that facilitate specialization on cultivation include increased size, deeper and fuller grooves, use of dense material, and extensive edge grinding.

The data set with which I am most familiar is from the Mimbres Valley, southwestern New Mexico. The largest axes found are those made during the Classic Mimbres period, when cultivation was most extensive. These were made almost exclusively of greenstone, an extremely hard, dense metamorphic rock local to the area. Greenstone flake debitage is most frequent in deposits from this period. In the Mimbres, these indicators of emphasis on reliability of axe design imply increased use of stone axes. Other evidence confirms that the increase was during a period of intensive and extensive plant cultivation (Minnis 1985).

Axes occur in household contexts and burials. Boydston (1989), in reference to ground stone axes in New Guinea, notes the interrelationship between the utilitarian and social values of highly specialized tools. Not only are they efficient, but well-made objects can become symbols of power and rank and gain value as exchange items. In an exchange system they may serve as a way to store wealth that can be traded for food in times of need (Boydston 1989:75–76).

Food grinding technology should also change with a shift toward specialized dependence on cultigens, particularly maize. Grinding tool and facility designs should be improved in use-efficiency to decrease the per unit time spent processing maize (Hard 1990; Lancaster 1983; Mauldin 1991). In addition, grinding tool design should have been reliable, to decrease breakage potential. Robert Hard (1986, 1990) has written extensively on grinding efficiency, discussing the advantage of staged grinding, mano form, and mano length. Multistage grinding is especially efficient; this is achieved through manufacture of grinding stones of different textures. In the Mimbres area these are vesicular basalt for the coarse grind, rhyolite for the medium grind, and sandstone for the fine grind. This kind of three-stage grinding is common in the late pueblo periods. Hard has also argued that finger grooves in manos make them easier to grasp, improving grinding efficiency. Finally, mano length is related to grinding efficiency, as it is one way to increase the surface area of manos. Hard has examined a variety of ethnographic cases of maize grinding and found a close correlation between the amount of maize consumed and mean mano length (Hard 1990). I would argue that to increase the reliability of these grinding tools, thickness is also important. Thicker manos and metates have a longer potential use-life because they can be repecked over a longer period and break less often. Christenson (personal communication 1995) suggests that finer grinding resulting from multistaged processing reduces cooking time and correlates with fuel shortage. Thus, this grinding technology is more efficient but only in the "context of labor management."

A third kind of tool may have facilitated specialization on cultivation. Vessels designed to cook slowly for a long time would be best suited to maize cooking (Braun 1983). They would also allow household cooks to engage in other activities without much concern for damage to cooking food. The work of Skibo, Schiffer, Blinman, and Wilson (1991) on performance characteristics of cooking vessels will allow us to recognize such changes in ceramic technology. Arnold (1985:144) has noted that spherical forms and those that are fired at relatively high temperatures are more resistant to thermal shock and therefore should be less likely to break in use. Concommitant change in grinding and cooking technology is consistent with Christenson's view of technological change in maize processing.

Because cultivators also hunted and gathered, they may have specialized by restricting the plant and animal species they exploit. This strategy should be evident in the use of facilities (water channeling devices, storage features) that increase the productivity and storage potential of wild resources and in the production of reliable, efficient, standardized tools and weapons. These are considered below.

HUNTING. If cultivators responded to stress by specializing on one or a few game species, facilities or methods for increasing the productivity of the target resources should have been used. Facilities that improve the predictability of an accurate hit reduce the chance of loss (Myers 1989:87–90). These may be tended facilities, which work for hunters when they are present, or untended facilities, which work when the hunters are not present. Hunting blinds (tended facilities) allow hunters to reduce the distance between themselves and their game. Channeling devices reduce the spatial distribution of game, both decreasing the distance between game and hunter and reducing escape potential for the game. These tended facilities are effective for group hunting by driving game. While group hunting does not require tended facilities, capture efficiency is improved through their use (see Kaplan, Hill, and Huertado [1990] and Hill and Hawkes [1983] for a discussion of Ache group hunting, Baksh and Johnson [1990] for a discussion of group hunting by the Machiguenga, and Wilkie and Curran [1991] on Mbuti net and bow hunting). Traps, which are untended facilities, multiply the simultaneous capture opportunities of hunters (Winterhalder 1990:86). Traps are valuable in contexts where the spatial variability of resources is high (Torrence 1989:60) and there are scheduling conflicts among valuable resources. In the Southwest, artiodactyls, which are a primary medium-to-large game resource, are most concentrated and in best condition in the fall, which is also when cultigens must be harvested.

Hunting blinds and drives are visible archaeologically as alignments of rock, but are difficult to date. Nets and most trap components are less visible, because they are most often made from perishable material. Artifact collections from dry shelters and caves, where perishables are preserved, may be useful for understanding technological strategies of hunting. A detailed account of the variety of facility-aided hunting techniques employed in North America is provided by Anell (1969), and may be an excellent source for beginning to identify facility use in the Southwest.

The tools and weapons of specialized hunting are expected to be reliable, use-efficient and standardized. As discussed above, reliability may be achieved by overdesigning the haft. Secure, thick hafting elements on stone points insure against point breakage and loss. Christenson (1987:145–147) argues that hafting is made secure on projectile points by maximizing contact of the point with the seizing. Wide stems and wide bases provide this contact. Side notching provides a secure hafting element, as has been noted for Southwestern projectile points by Jelinek (1967:110) and Kidder (1932:22). Also, stem length adds security to a haft (Christenson 1987:147; Jelinek 1967:11). Tool thickness and basal grinding of the stem improve reliability. The thickness of a tool contributes to its reliability because breakage is less likely to occur during use. Basal grinding of projectile points protects the shaft from splitting on impact by contact with the sharp base of the point (Christenson 1987:148).

These investments in stone point design may result in production of larger points that can be reworked. Investments in production can be recovered by reusing and reworking a tool (e.g., Binford 1979; Nelson 1991). Production investment also suggests selection of durable, knappable material. As Nelson (1981) and Horsfall

(1987) discuss, these characteristics tend to be inversely related. The most brittle material, obsidian, is least durable and the least workable materials, some metamorphic and igneous rocks are most durable. Goodyear (1989) has noted that cryptocrystalline material is well suited to projectile point manufacture and use. It is easily knapped and relatively durable.

Recovering investment in production through tool reuse assumes that tools can be recovered after use. Flenniken and Raymond (1986) note the high breakage rate for projectile points, but also observe through replication experiments that rejuvenation is considerably more economical in time than manufacture. Experimental manufacturing of stone points required more than ten times the amount of time of rejuvenation (Flenniken and Raymond 1986:604). Though both require minutes rather than hours, frequent replacement does consume considerable time. In addition, Flenniken's figures do not include the amount of time required to rehaft new point tips or to replace shafts and foreshafts. Thus, efforts should be made to recover tools which have been made with extra production efforts in time and material.

Recovery of projectile points for rejuvenation and reuse has implications for the form and context of their use. Sloping shoulders would allow removal from an animal more readily than barbed shoulders. Secure hafts and thick cross section would protect against breakage. The hunting context in which points can be recovered is limited to close shots. Distance shots would be unlikely strategies for use of points designed for recovery and reuse. Animal tracking, in which it is important for the point to remain in the animal to exacerbate bleeding, would not be a strategy suitable for recovering points.

In short, relatively large, thick, sloping shoulder, wide-based and wide-stemmed points which are not made from glasses or dense rock would suit a specialized hunting strategy.

GATHERING. It is difficult to imagine that specialization on one or a few wild plant resources would be an effective risk-reducing strategy in the Southwest. Two resources may have been emphasized: pinyon in the uplands, and agave in the desert. The productivity of both resources can be improved through attention to their growth needs. Manipulation of wild plants to increase their productivity by improving the growth environment will decrease the time and energy spent in acquisition and increase yield. Minnis (1985b) suggests that strategies to manipulate plant productivity were employed prior to full-scale cultivation in the Southwest.[2] Fish et al. (1986) describe the manipulation of agave in southern Arizona, visible in the stone mounds formed in agave gardens. The location of productive pinyon is somewhat unpredictable because of the long time span between years of cone production. Wills (1988) has argued that settlement in the uplands would aid in monitoring and controlling access to pinyon groves. Pinyon and agave may have been important staples or supplements because their location is more predictable

[2] See also the chapter by Alan Sullivan in this volume.

than is that of game. However, in portions of the Southwest that receive most of their moisture through summer thunderstorms, locational predictability of adequately watered plant foods is low. For a specialist focusing on one or a few plant resources, this is further reason to manipulate their productivity and control access to them.

Specialized focus on wild plants is difficult to identify through storage facilities and tools. The storage needs of many plants are similar, so that specialization on a few wild plant resources is not distinguishable on the basis of increased storage capacity among agriculturalists. However, agave roasting pits are a unique feature that would aid in recognizing agave processing. The tools employed for gathering would rarely be visible archaeologically because so many are made from portable material, but agave processing knives may be identifiable (Fish et al. 1986).

Economic specialization, thus, may have been implemented through the manufacture and use of a variety of reliable, specialized, efficient tool forms as well as a variety of facilities for improving productivity and storage potential of food. While considerable investment in tool manufacture may have been required, it would have resulted in improved performance qualities.

DIVERSIFICATION OF RESOURCES

Diversification of resources involves an increase in the number of kinds of food acquired and consumed. With this subsistence strategy, the effect of variance in any one resource is reduced by dependence on many resources. Another way of viewing diversification is that variance in productivity of any one resource area is reduced by use of a variety of resource areas, spreading risk (Reid [1978] referencing Watt [1972:75–76]) and taking advantage of varied conditions (Bamforth and Bleed 1991; Winterhalder 1989).

During periods in Southwestern prehistory when agriculture was extensively practiced, two kinds of diversity may have developed. The first involves adding crops to the range of cultigens. This I have subsumed under specialization because of the focus on one class of resources. Ford (1981) discusses the relatively late introduction of beans, and other cultigens, at a time when maize formed a major part of the prehistoric diet. With this kind of change, a wider variety of cooking and storage vessels might have been needed, or designs that work well for cooking a range of kinds of food may have been advantageous. Attention to the performance characteristics of vessels (Schiffer and Skibo 1987) and efforts to evaluate multifunctional vessels would assist in addressing these points.

The second kind of diversity involves expanding on the wild resources that are part of the subsistence inventory. While wild resources were exploited throughout the agricultural prehistory of the Southwest, dependence on greater variety may have occurred as specialization in cultivation was not providing adequate food (Reid 1973). For example, Graves et al. (1982) discuss an increase in the variety of game, especially deer, during unfavorable seasons of occupation at Grasshopper

Pueblo. I have discussed the same for the Classic Mimbres period in the Mimbres Valley (Nelson 1981). While failure of the specialized agricultural economy may have resulted in abandonment of regions (Cordell 1975; Fish, et al 1989), it also may have been responded to with addition of a more systematic dependence on wild resources, among other strategies. The development of agriculture may, in fact, have made diversifying the use of wild resources more feasible. Szuter (1995) argues that animal populations increase in size and diversity with the introduction of cultigens to an area.

Diversification of wild resource use can be accomplished by adding onto the investment in cultivation, increasing the labor requirements of all members of society, and remaining in large aggregated settlements. Alternatively, it may be accomplished by dispersion into smaller, more mobile groups that invest less in cultivation. While the settlement and social strategies for diversification are of interest, there is not room to consider them in this chapter. I assume for discussion that the expanded range of wild resources was added to an intensive agricultural base, an assumption appropriate to the context of the agriculturally based, aggregated settlements of late puebloan prehistory.

Hunting or gathering activities may compete with cultivation unless the resources used are those attracted to cultivated fields. Under conditions of low and spatially variable resources density, common in the Southwest, time invested in locating wild resources can be considerable especially for mobile resources such as game (see Torrence [1983, 1989] for detailed discussion of time stress). Cultivators can approach the addition of hunting or gathering to their inventory of activities by specializing or by diversifying. Diversifying strategies favor variety over intensity and movement over spatial focus. More time is spent searching for food than in a specialized subsistence strategy (Bamforth and Bleed 1991; Winterhalder 1989) though processing for storage may involve less time. Facilities that decrease search time would be appropriate to this economic strategy. Further, the potential for uncertainty regarding the specific range of resources that may be encountered on any foraging trip is increased by the diversity of resources sought and areas exploited. However, uncertainty need not imply stress if technology is designed to respond to uncertainty. Versatile, flexible, portable tools and weapons are designed for variety and uncertainty.

Facilities should be varied, but not in the same way as is expected for specialists. The specialist's assemblage of facilities is designed to approach one or a few resources with various strategies for insuring capture and productivity. The diversity of generalist's facilities results from employing a variety of ways to exploit different resources. For example, the generalist might employ different kinds of traps to encompass the variety of ways that different animals can be trapped, while the specialist might employ different traps designed to capture the same animal in different contexts. For a generalizing hunting strategy, facilities should be effective for capturing game of a variety of sizes and prey responses (scatter, cluster, freeze, fight).

Tools and weapons are expected to be generalized rather than diverse because of the demand for portability imposed by exploitation of a variety of resources. Numerous specialized tools for each exploited resource would comprise an overwhelmingly bulky collection. Tools and weapons that are general-purpose can be used effectively to respond to the variety of needs encountered in exploiting a wide range of plant and animal resources. As noted above, the potential for uncertainty in the timing of specific tool needs can be addressed with multipurpose tools (Jochim 1989; Nelson 1991). Multipurpose design can be achieved through tool flexibility or tool versatility (Nelson 1991:70–73). Emphasizing flexible design insures that a tool can be reshaped easily to suit a variety of needs; versatility accomplished the same end with a generalized form.

Large bifacial cores are both versatile and flexible. A variety of flake forms can be produced from them and their generalized form can itself be used in a range of cutting, chopping, and scraping tasks (Binford 1979: Kelly 1988; Morrow 1987; Nelson 1991:72). This would be an excellent tool kit for butchering and for generalized foraging, as it contains both cutting and chopping elements. The flexible, versatile bifacial core is also easily maintained. Torrence (1989:63) notes that maintainability is important in contexts where there is variation in the number of times per foraging trip that a tool may be needed. Bifacial cores do occur with some regularity on Archaic sites in the Southwest, but to my knowledge, they are rare in assemblages from later periods, suggesting that generalized tool design may have been rare in the post-Archaic Southwest. Cameron (1987:113) records the presence of bifacial cores in Basketmaker II, Early Ceramic and Late Ceramic assemblages from Black Mesa as 8 to 11 percent of the core samples. More attention to identification of bifacial cores and examination of changes in their proportional frequency may aid in identifying the application of generalized technological strategies.

Flexibility can also be achieved within tool kits by employing a modular design (such as the Swiss Army knife). Using replaceable foreshafts, different tool ends can be fitted to the same handle, but Ellis (1994:22) notes that the foreshaft adds a point of weakness. In the archaeological record, use of foreshafts might be evident in the same stem and base form on tools with different shapes of blade or functional element (Nelson 1991). I am not aware of any attempt to evaluate the use of this strategy in the prehistory of the Southwest.

While a shift to generalized tool forms is expected in the tool assemblage of resource generalists, hunting weaponry must always have some element of reliability. Failure of a weapon to perform quickly can result in loss of the game, which is not true for acquisition of plant resources (Torrence 1989). A design for stone-tipped weapons that incorporates a degree of both reliability and versatility emphasizes manufacture of multiple, parallel duplicates. Such weaponry might include numerous arrows of similar size, all of which can be quickly projected from the same bow. These arrows may be smaller than those of specialized hunters because accomplishing versatility and reliability through access to numerous duplicate arrows or foreshafts could be cumbersome if each projectile was large. If arrow points are small, then shafts can be small and light (if fletched—see Christenson 1987; Parks n.d.), facilitating transport of many arrows (see Nelson 1991:73–76 for discussion of transportability of tools and toolkits). Ellis (1994) has shown in an extensive cross-cultural study of hunting technology that stone tips of all sizes are effectively used to take animals the size of those hunted in the Southwest. However, Christenson (1987) and Parks (n.d.) argue, all else being equal, small points have less penetration potential for slowing and killing large game, especially for distance shots (see also Browne 1940). Loss of penetration potential can be compensated by improving the draw power of the bow, which determines initial velocity of an arrow, and by modifying the form so that it cuts broadly into and stays in the animal, both of which exacerbate hemorrhaging (Christenson 1985; Parks n.d.). Barbs contribute to the killing power of a point by maximizing blade length and causing the point to remain lodged in the animal (Christenson 1987:145–148; Jelinek 1967:110). In addition, barbed points are difficult to remove because the angle of the shoulder tears at tissue as it is pulled out. This is especially important when hunters are tracking game after the animal is shot.

Designing a point to stay in an animal has a cost. The point and the shaft or foreshaft are lost. While production of stone points is not very time consuming, manufacturing well-balanced, streamlined shafts is much more so (Keeley 1982; see Flenniken [1986] for the differential in time spent rejuvenating as opposed to manufacturing new stone points). However, loose hafting can cause the point to separate from the shaft, leaving the point in the animal to exacerbate hemorrhaging and allowing the shaft to be retrieved. Ellis (1994:13) cites a study of California Yurok points in which barbed points were loosely hafted and remained in the prey when the shaft was removed. Points with narrow, short, parallel, or contracting stems have relatively loose hafts (Christenson 1987:145–148). Simple triangular points are also less securely hafted than are notched points (Kidder 1932:22). These points, designed to be lost in the hunted animals, should be manufactured with little investment in formal shaping beyond the production of forms with the characteristics described above. The small triangular points so common in the late prehistory of the Desert West may be just such "throw away" projectile tips.

Thus, the diversifying response to resource stress that requires hunting a range of animals in varied contexts would be effectively implemented with a reliable, versatile, and portable tool kit. Projectile points should be small, loosely hafted, with barbed shoulders and made with less production investment than are the reliable, specialized weapons. Flenniken's (1980:29) experimental manufacturing of projectile points leads him to the conclusion that small points made on flakes, using few reduction steps, are the least costly, measured in production time.

SUMMARY

I have described contrasts in construction of facilities and in design of tools and weapons. For resource specialists, facilities designed specifically to enhance the productivity, spatial and temporal predictability, and storage life of a narrow range of target resources are expected. In addition, the design of tools and weapons should emphasize reliability, use-efficiency, and specialization. This design combination could result in a diverse tool kit of specialized forms. For resource generalists, facilities, tools, and weapons should be generalized to suit a wide range of possible resources and be responsive to varied, and potentially uncertain, future conditions. Tool and weapon designs should combine reliability, versatility, flexibility, and portability.

Under conditions of resource plenty the correlation between economic and technological strategies may not apply. However, where stress and risk are occurring, tools and facilities are expected to approach optimal design strategies for the particular stress or risk responses employed. The hunting weaponry of late prehistoric occupants in the Southwest is examined, next, with these contrasts in mind.

APPLICATION OF THE TECHNOLOGICAL STRATEGIES APPROACH

Technological strategies can be used to reduce the probability of subsistence risk and stress. The greater the risk, the more costly technological failure, which increases the benefit of designing tools and facilities for most effective capture and processing. Under stress, not only should tools maximize the probability of resource acquisition, but they should be designed so that their production and repair competes as little as possible with other activities crucial to reducing subsistence stress. The tool and facility designs that are most effective are determined by the social and economic strategies employed by people to respond to risk and stress. Thus, by examining the material remains of technology within the historical context of their use, ideas about human responses to stress can be proposed and evaluated. I discuss one issue of concern to Southwestern archaeologists and how technological analysis can contribute to resolution of this issue.

Southwestern archaeologists have long debated the causes of large village and regional abandonment. Many explanations include resource scarcity as a contributor to conditions that led to abandonment of large villages. If prehistoric occupants of large villages in the late prehistoric phases of Southwestern prehistory (post A.D. 900) were experiencing stress, they may have adopted specialized or diversified economic strategies for addressing that stress or the risks of future shortfalls. An assessment of the technological strategies employed prior to abandonments can contribute to our understanding of the extent of subsistence stress and the possible roles of different strategies in ameliorating stress. If cultivators first responded to subsistence stress by specializing on cultigens, as seems to be the case in many regions of the Southwest, we should expect to see those features that indicate efforts to increase productivity of plant resources, and efficiency in processing and storage. Abundant evidence of water control and storage have been documented in various areas in the late ceramic periods. By these late prehistoric periods, grinding technology was specialized and staged in many areas. Attention to axe design and the forms of cooking and storage vessels would supplement these data.

Of further interest is a consideration of whether cultivators, prior to abandonment of large sites or regions, modified their subsistence efforts toward greater dependence on wild resources in systematic, technologically effective ways, either by specializing or diversifying. If cultivators added plant and/or animal resources onto their heavy investment in cultivation as a risk reduction or stress response, the technology employed would need to be cost effective. While exploitation of wild plant resources is difficult to see in stone technology, hunting strategies are more easily studied. Ellis' (1994:6–7) cross-cultural study of hunting indicates that stone-tipped weapons are used nearly exclusively on large game (>40 kg.) and on humans. Thus, the stone projectile data analyzed in this chapter are assumed to be the products of hunting large game, rather than rabbits, birds, and rodents. Their use on humans is not assessed.

The specialist strategy would focus on one or a few animals taken in a narrower range of contexts than the generalist hunting strategy. If a diverse range of game were depended upon, we should see a versatile, portable hunting weaponry that is effective on a range of animals with varied prey responses. Alternatively, if cultivators took a specialized focus in hunting, a weapon design with more emphasis on reliability and specialized form, suited to hunting large game in a narrower range of contexts, should be evident. If we do not find either combination of design options predominating, it is possible to conclude that effective hunting strategies were not risk reducers or that stress was not occurring.

The responses of prehistoric cultivators to risk and stress across the Southwest would have varied, given the diversity of environmental and social conditions. Environments vary with regard to density and diversity of large game. Social mechanisms provide access to game or other resources through trade and exchange (e.g., Graves, et al. 1982; Spielmann 1983). However, to begin an exploration of the role of hunting technology in economic responses to stress, I have gathered data on projectile points from late occupations in several locations. This initial analysis illustrates the value of the technological approach, but is inconclusive due to the limitations of published data.

I begin by assuming that cultivators living in large aggregated settlements experienced subsistence stress late in the occupation of regions. The role of hunting as a stress response is considered for these regions. There are many other options for cultivators responding to subsistence stress. Specialization in cultivation (which may have preceded diversification), movement to new areas (which may have followed failed diversification efforts), and trade for game resources are only a few. All of these occurred in the Southwest, but I am looking at whether we can suggest that hunting specialization or diversification played a part in the late puebloan strategies of addressing subsistence risks.

Projectile point data from Chaco Canyon (Lekson 1985), Gran Quivira (Judi Green 1985), Tijeras Pueblo (Blevins 1974; Cordell personal communication, 1992),

Black Mesa (Cameron 1987; Christenson 1987), and Mimbres (Nelson 1986) are discussed (Figure 1). These are regions for which there are adequate descriptions of sizable projectile point collections. Samples from the different regions are not contemporaneous; I have selected from contexts in which cultivation was well developed and from periods just prior to large village or regional abandonments, or major social change.

Unfortunately, the data on projectile points are recorded differently among these various research areas. Many of the variables important to evaluating the design dimensions (reliability, versatility) most emphasized during any period are not consistently recorded or their associations are not analyzed. These include length, thickness, and workmanship characteristics. The only variables consistently recorded and examined in association are shoulder angle and stem or haft form. Shoulder angle affects the staying potential of points after penetration; barbed shoulders fix in the animal, while sloping shouldered points do not fix in the animal (Figure 2). Fixing of a point in an animal increases the hemorrhaging of the wound,

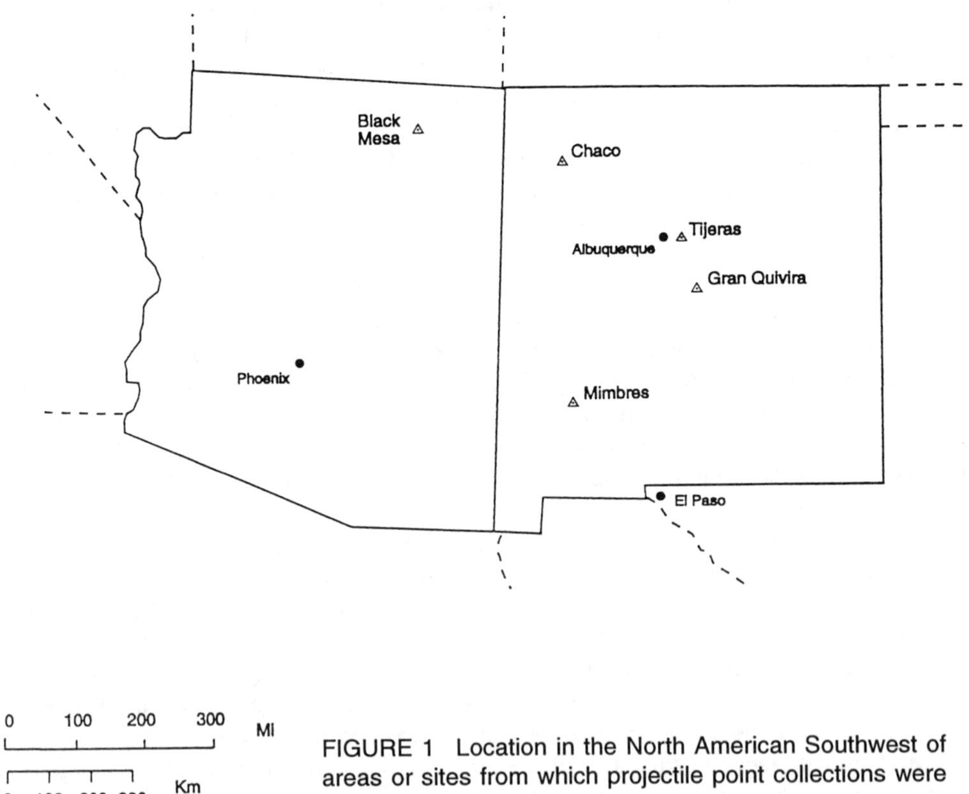

FIGURE 1 Location in the North American Southwest of areas or sites from which projectile point collections were examined.

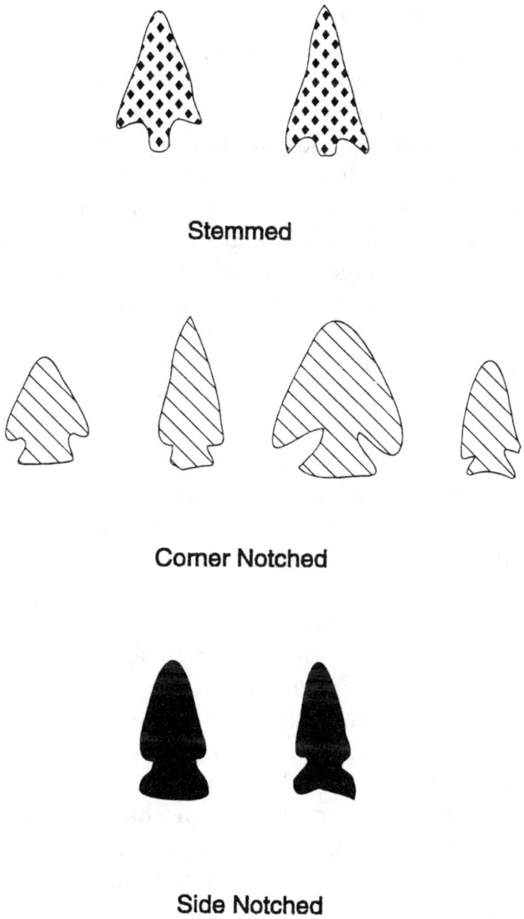

FIGURE 2 The three primary stone projectile point forms in post-Archaic collections in the North American Southwest.

which increases the likelihood of a successful kill. This is especially important when tracking game and taking distance shots. For close range shots on driven game where the animal can be shot multiple times, the point need not fix in the animal. If sloping shouldered points do not fix in animals, some may be recovered, rejuvenated, and reused. But contexts in which points would be recoverable are limited.

Stem shape also contributes to the fixing of points in animals and potential recovery of the shaft. Constricted stems could more easily be separated from the shaft than could side-notched, wide-base points, potentially allowing the shaft to fall away from the animal. However, if foreshafts were used, the shaft may be recoverable regardless of point form.

Data from five regions in the North American Southwest show patterned difference in hunting technology during the late occupations of large, aggregated village settlements (Figure 3). In some areas, a single point form with sloping shoulders and secure haft (deep side notches) dominates the projectile point collections. These are points that would not fix well in the animal, but would be potentially recoverable. Their dominance correlates with an emphasis on hunting animals that herd. In other areas, no single point form dominates. These patterns are discussed below.

Most studies of projectile points in the Southwest employ typology rather than technological analysis of the occurrence of single characteristics (but see Jelinek [1967] and Christenson [1987]). In 1967, Jelinek described a range of formal types for the Middle Pecos; these are used to a greater or lesser extent in most studies. Fortunately they include variables of hafting element and barbing. The most common post-Archaic combinations (Jelinek 1967:104–106) include: corner notched with parallel to contracting stems and barbed shoulders, often referred to as stemmed; corner notched with expanding stem, often with barbed shoulders; and side notched, often without barbs (Figure 2). Other possible combinations of stem shape, notch position, and shoulder shape are possible and do occur in the Pecos collection as well as in others (Jelinek 1967: Figure 15). I focus primarily on these three types, grouping data from reports as necessary.

Jelinek demonstrates that the three primary forms are good time markers in the Middle Pecos with the stemmed points occurring early (Early 18 Mile Phase) and the side-notched points dominating late contexts (Late McKenzie and Post-McKenzie phases). This sequence occurs for the Chaco sites (Lekson 1985) and is suggested at Tijeras (Blevins 1974), but is not applicable to the Mimbres and Black Mesa areas. Whether or not point form changes in a patterned way over time, the differences indicate modifications in hafting (Jelinek 1967; Keeley 1982) and hunting strategies (Nelson 1986). The three forms are

1. *Stemmed with barbed shoulder*: Point designed to remain in animal but release from the shaft (jam haft).
2. *Corner-notched, expanding stem, barbed shoulders*: Point designed to remain in animal, but also to remain securely hafted to the shaft.
3. *Side-notched with sloping shoulder*: Point designed to be easily removed from animal and to remain securely hafted.

As noted above, the *stemmed, barbed point* is a generalized design for bringing down game with a variety of hunting techniques (communal and individual, drive and stalk). It is effective for a diverse range of hunting strategies, and is also designed to potentially save on the cost of replacing the shaft after each successful shot. Note that shafts may be broken when penetration occurs, but this is not inevitable as it is with more securely hafted points and shafts. The *corner-notched expanding stem point* is equally versatile, but the shaft would be lost at each hit, because of the secure haft afforded by the wide stem. The *side-notched forms* with secure hafts and sloping shoulders are best suited for conditions in which the game need not be tracked after shot, because they are not designed to fix in the animal. This is a narrower range of hunting conditions than is expected for a resource generalist. If reuse of these latter points was anticipated, they should be thicker and initially larger than those used by the generalized hunter. A wide range of base forms may also indicate that a diverse range of game were taken or at least that diverse hafting or hunting strategies were employed. The cost of such tool kit diversity needs to be evaluated.

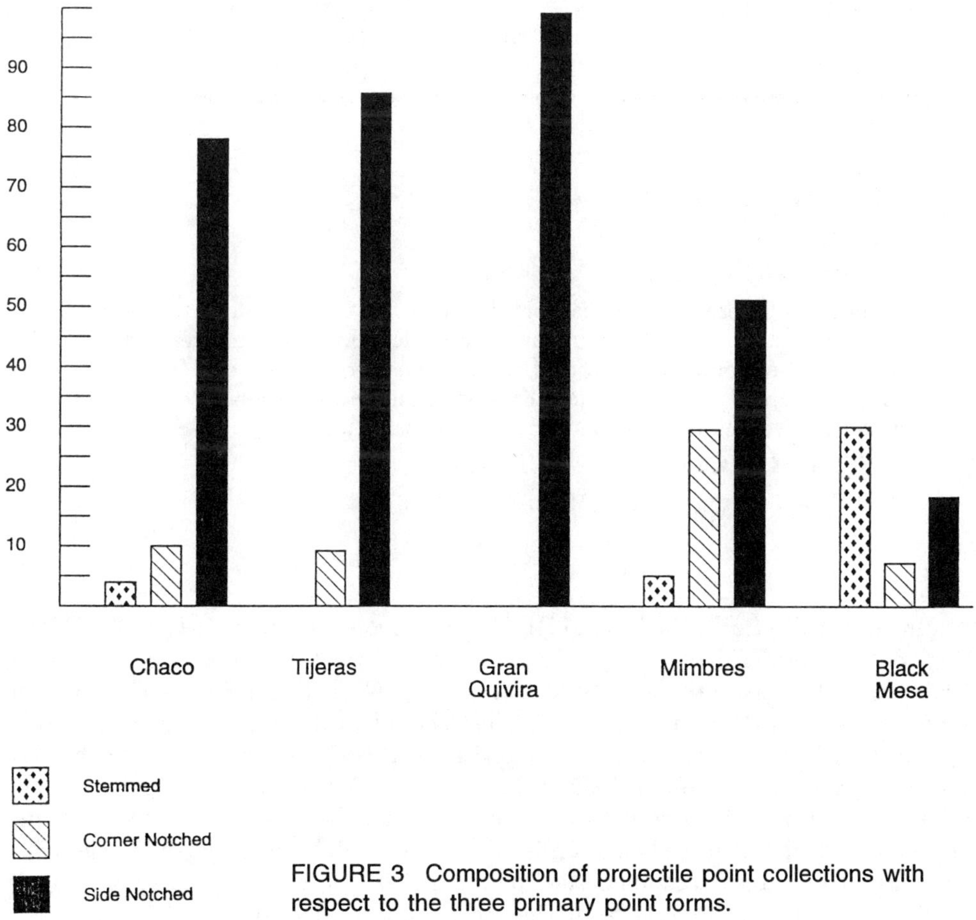

FIGURE 3 Composition of projectile point collections with respect to the three primary point forms.

TABLE 1 Proportional frequencies of the three primary projectile point forms on late prehistoric sites in the Southwest.

SITE/REGION	POINT FORMS				
	Stemmed	Corner-notched	Side-notched	Other*	N
Chaco Canyon[1]	3.2	17.4	79.4	0	nd
Tijeras Pueblo[2]	nd	nd	87.8	12.2	147
Gran Quivira[3]	0	0	100.0	0	nd
Mimbres[4]	nd	30.0	55.0	15.0	284
Black Mesa[5]	29.7**	5.4***	18.9	45.9	37

1 From Lekson (1985: Table X.2, X.3).

2 From Blevins (1974: Table 2).

3 From Green (1985: Figure 5).

4 From Nelson (1986: Table 8.23).

5 From Cameron (1987: Table 4-36).

6 *For the Tijeras Pueblo sample, forms other than side-notched are not summarized separately in the data tables. For the Mimbres sample, this includes stemmed and other point forms. For the Black Mesa sample, "other" refers to generalized stemmed and corner-notched points, as well as unidentified and fragmentary forms.

7 **Type 54 in Cameron (1987: Table 4-36).

8 ***Type 55 in Cameron (1987: Table 4-36).

In three collections, the Pueblo III Chaco Canyon collection (Lekson 1985), the Pueblo IV Tijeras Pueblo sample (Blevins 1974, Cordell personal communication 1992), and the sixteenth to seventeenth century Gran Quivira collection (Green 1985), side-notched points without barbed shoulders dominate (Figure 3; Table 1). In the Chaco collection, these comprise nearly 80 percent of the Pueblo III notched arrow points (Lekson 1985: Table 1, Figure x.1). Within the Tijeras Pueblo collection over half (55 percent) of the points are notched; among these 88 percent are side-notched, but no information is available regarding the angle of the shoulder (Blevins 1974: Table 2). Side-notched points decrease in the collection from a high of 95 percent in the sample from the second phase to a low of 81.5 percent in the third and final phase of site occupation. For the Gran Quivira collection, only two point forms are reported, triangles and side-notched; the latter have sloping shoulders in most cases (Green 1985: Figure 5). These securely hafted, specialized forms are designed for reuse.

If these forms were made with the intention that they be reused, they should be thicker and heavier and have more reworking than do the point forms from earlier occupations. For the Chaco collection, Lekson (1985:18) notes no significant differences in metric variables of size among the different points. Reworking on the side-notched points is not discussed. Data are not currently available on these characteristic for the Tijeras and Gran Quivira samples, although in the Tijeras collection, there is little difference in other aspects of point form over the three occupations described by Blevins (1974: Table 2).

I have argued above that a technology designed for economic specialization should have more standardized forms. The coefficient of variation for metric variables is one index of standardization. For the five metric variables reported by Lekson for the projectile points in Chaco collections, the side-notched points have the smallest coefficient of variation of the three point forms, for all but one of the measures, suggesting greater standardization especially for shoulder and base width and stem diameter (Table 2 from Lekson [1985: Table x.3]).

At both Chaco and Gran Quivira, it has been argued, prehistoric people obtained meat through trade (Spielmann 1983; Lekson [1985] referencing Aikens [1982]). The projectile point collections indicate that meat may have been acquired also through specialized group hunting, a strategy that may have developed to insure some access to meat or important nutrients from game (see Spielmann, this volume) during periods of subsistence stress. Herding animals may be more frequent in these areas than in others, which should be evident in the faunal samples. Spielmann (1995) has suggested that large herding animals are more prevalent and a more dominant part of the faunal assemblage at late prehistoric sites in the eastern portion of the Southwest than in the western portion. At Gran Quivira, the faunal assemblage is dominated by antelope remains (Spielmann, personal communication 1992)

In collections from the Classic Mimbres occupation (eleventh and twelfth centuries) of the Mimbres Valley (Nelson 1986), the projectile point collections are more diverse than are those from the Chaco, Tijeras, and Gran Quivira areas. Side-notched points with sloping shoulders, so common in the late Chaco Canyon, Tijeras, and Gran Quivira collections, form 55 percent of the collection from the Classic Mimbres period (Figure 3, Table 1). They show little production investment. Most are not pressure flaked over the entire face on both sides, their notches are shallow, and they are small, contrary to expectations about reliable hunting weaponry. A wide range of other barbed and nonbarbed forms is present, although stemmed points are rare. The Classic period sample has narrower hafts than are evident in samples from earlier occupations, but these occur on side-notched points, which may indicate that the notches are deep, producing a narrow neck. The Mimbreños did not consistently use a strategy of producing points that would detach from their shaft (straight-stemmed points) and remain in game (barbed-shoulder points). In fact, barbing consistently occurs on points with an expanding stem; these form 30 percent of the Classic Mimbres collection, down from earlier periods.

TABLE 2 Dimensions of point forms from Chacoan sites (from Lekson 1985, Table x.3).

	Stemmed	Corner-Notched	Side-Notched
Blade length			
$\bar{X}$	20.54	21.08	19.79
s.d.[1]	6.42	6.38	5.34
N	55	154	252
V[2]	.31	.30	.27
Base length			
$\bar{X}$	4.91	4.67	5.46
s.d.	2.10	1.15	1.51
N	55	161	268
V	.43	.25	.28
Shoulder width			
$\bar{X}$	13.73	12.71	11.78
s.d.	3.94	2.79	1.71
N	60	163	268
V	.29	.22	.15
Minimum stem diameter			
$\bar{X}$	5:35	6.72	7.47
s.d.	1.66	1.52	1.32
N	60	164	272
V	.31	.23	.18
Base width			
$\bar{X}$	5.69	10.78	12.52
s.d.	2.66	2.23	2.16
N	58	161	266
V	.47	.21	.17

[1] V = coefficient of variation

[2] s.d. = standard deviation

The projectile point collection from the Mimbres area does not indicate a strong emphasis on one hunting strategy. Although generalized hunting is indicated by the range of shoulder and haft forms, the tool kit is not designed to maximize versatility and minimize production time and effort. It is possible that hunting was not a dependable subsistence activity for hedging against the threat of food shortfalls prior to the abandonment of the large sites in the Mimbres Valley. Analysis of faunal material indicates that a wide range of game was taken during the Classic Mimbres period, with large herding animals less common than in other areas to the north (such as Chaco). Hunting by individually stalking game rather than hunting herding animals may not have been effective for substantially diminishing subsistence risk over the long term in the arid to semiarid Southwest.

The hunting weaponry from Black Mesa is included as the last sample. Although no large aggregated settlements occurred in the late ceramic period (A.D. 1050–1150) in the study area of the northern mesa, the data on projectile points are the most complete of any available. Two kinds of data are available. Cameron (1987) conducted a typological analysis, most similar to those described for the other collections, that merges samples from all ceramic periods (A.D. 800–1150). Christenson's (1987) analysis separates the collection from the late ceramic period in a technological analysis. Christenson found little change in projectile point form within the ceramic periods, except for a decline in barbing, which may indicate a decline in emphasis on point forms designed to stay in the animals. Slight increases in haft length and base width suggest manufacture of more secure hafts and perhaps intended reuse, although Christenson finds no significant change in incidence of retouch. The decline in barbing, increase in haft length, and increase in base width all indicate a trend toward greater emphasis on reliability and reuse, but not a significant or substantial shift.

In her analysis of overall point form, Cameron grouped all points from the ceramic periods and contrasted them with Basketmaker II and Archaic collections. A variety of hunting strategies is indicated for the ceramic periods, as was the case for the Classic Mimbres collection, but the predominant point forms are different (Figure 3, Table 1). Side-notched points form nearly 19 percent of the ceramic period sample, points with straight stems and barbed shoulders form almost 30 percent. These latter are the lightest and thinnest, as expected for an effective generalized design, but they have the greatest investment in workmanship (Cameron 1987:130), which is contrary to expectations for a point designed to lodge in an animal and be lost. Social demands on point form and utilitarian demands on their use as weapons against humans may influence workmanship. The side-notched points with deep notches also show high levels of workmanship (Cameron 1987:130), which is as I expect for a reliable design.

The point forms from the Ceramic period collections on Black Mesa indicate a variety of hunting and manufacturing strategies, with little change over 350 years of occupation. This is consistent with the lack of change in use of faunal resources (Leonard 1986). Approximately one-third of the ceramic period points are designed to be versatile (barbed, straight stemmed, small points). However, the lack of change over time in game selection and hunting technology indicates that hunting was not emphasized *exclusively* in the late ceramic period to reduce subsistence stress or that subsistence stress was not occurring on a regular basis. Perhaps resource stress was not the cause of regional abandonment on Black Mesa.

Among the five collections examined, three (Chaco, Tijeras, and Gran Quivira) are dominated by side-notched points with sloping shoulders, which are suited to a narrower range of hunting conditions than the barbed, narrow-stemmed points. They are more specialized but also potentially recoverable and reusable. There is some indication that these forms were more standardized than others in the Chaco Canyon sample. None of the collections is dominated by stemmed, barbed points, which are designed to maximize the versatility of the tool kit and minimize production, the technological strategy expected for a generalized, diverse hunting strategy responsive to stress. Two of the collections, Mimbres and Black Mesa, have a fairly wide range of point forms, which could have been used to capture a range of game, but is not an optimal tool kit for generalized hunting.

Lack of versatile point designs indicates that generalized hunting was not a good risk-reduction strategy in the Southwest. While people engaged in hunting a wide variety of game during some time periods, this practice may not have been effective as a risk-reducing strategy in times of food shortage or anticipated risk of shortage. Although Szuter (1995) argues that cultivation increases the diversity and number of game in an area, the game density is low in the arid to semiarid Southwest, perhaps too low to present a dependable resource for agriculturalists, unless large herding animals were taken. The preagricultural occupants of the Southwest depended to some degree on game, but their population levels were considerably lower than those of the late ceramic phases. It is interesting to note that Paleoindian and many Archaic projectiles have large, thick, securely hafted points with evidence of reuse. These reliable designs indicate the importance of taking large game during the preagricultural periods of Southwestern prehistory.

Stemmed points are common in some collections. For example, in the Basketmaker III—Pueblo I sample from Chaco—60 percent of the points are stemmed. If this is a versatile weapon design, its presence in this context may indicate that as agriculture developed, generalized hunting was effective for provisioning people while population levels were relatively low and aggregations considerably smaller than during the late prehistory of the Southwest. It would be interesting to examine other data from these periods to identify economic strategies and whether stress was occurring.

Simple, triangular, minimally retouched points are common in late prehistoric contexts. These are versatile, replaceable forms suited to generalized hunting strategies. I have not examined their contribution to hunting weaponry because they are not systematically included with the notched forms in many studies. Further research that includes all possible point forms is needed. In both the Tijeras and Gran Quivira collection their common occurrence is noted; perhaps these are the generalized hunting tips of the late periods.

The patterns identified in this study appear to support Jelinek's sequence for the Middle Pecos. The latest samples (Chaco, Tijeras, and Gran Quivira) are dominated by side-notched points and the earlier samples have more corner-notched and stemmed forms. However, this general diachronic observation is misleading for two reasons. First, all of the late samples are from the eastern portion of the Southwest, where, as Spielmann (1995) argues, large herding animals are more prevalent. The use of side-notched points during the late periods in these eastern areas may represent a stress-reducing strategy of specializing on group hunting of large animals in conjunction with an intensely agricultural economy. Such a strategy may not have been possible in the western portion of the Southwest, particularly in many parts of the Hohokam area. Thus, side-notched points should not dominate late period projectile point samples from sites in the western portion of the Southwest. This hypothesis needs further investigation.

Second, in the samples from Mimbres and Black Mesa, Jelinek's sequence does not fit. Stemmed points are not replaced by corner-notched points. In the Mimbres and Black Mesa collections, corner-notched points dominate the Early Pithouse and Basketmaker II samples, respectively (Nelson 1986; Cameron 1987), while in Chaco they do not become abundant until Pueblo II (Lekson 1985). Also, in the Black Mesa sample, corner-notched forms become less frequent from Basketmaker II to the ceramic periods, and stemmed forms become more frequent (Cameron 1987), which is opposite from the Middle Pecos sequence. The sequence of point forms in any region of the Southwest should not be expected to follow a single pattern. A number of conditioners may influence point design; among these, hunting strategies are an important consideration, as is their role in the overall economic strategies for addressing risk and food stress.

For this general discussion of the design of hunting weaponry, I have not systematically evaluated the differential availability of animal resources or material for producing weaponry across the Southwest. In some areas, game density, especially large game, may be low enough that any specialization on group hunting would not be possible. This may be true for the Jornada area or for portions of the Hohokam area. Similarly, there may be an absence of stone material in some portions of the Southwest such that conservation of tool material was important. These circumstances add additional dimensions to design considerations for prehistoric weaponry.

CONCLUSION

Technological strategies are the means by which humans use and modify their environment. They are, therefore, a sensitive index of the relationship between humans and resources. When groups are experiencing subsistence risk, changes in the ways in which resources are used may be expected. These changes should occur in the direction of improving the relationship between input and benefit. While technology may not always be organized toward optimizing effort, and tools and facilities may not always be designed for maximized returns, during times of stress technological organization and material design should approximate some optimal solutions to subsistence problems. "Forms of behavior which deal with the management of risk will be more sensitive to selective pressures and are therefore more likely to approach optimality" (Torrence 1989:4).

In this chapter, I have discussed economic and social responses to subsistence risk, and focused on the technological strategies that facilitate two kinds of economic responses: resource specialization and resource diversity. I have discussed aspects of technological strategy that would facilitate specialization and diversification among intensive cultivators who hunt and gather. These expectations about technology are based on the notion that technology can be instrumental in reducing the probability of resource loss. When the cost of that loss is high, as in situations of risk or stress, technological responses are expected to be most carefully designed to maximize returns (Bamforth and Bleed 1991).

For resource specialization, I have suggested that tool assemblages and facilities should be diverse as a result of the many techniques used to maximize returns from few resources. Further, I have argued that the design of tool and weapon form should emphasize reliability and use-efficiency; emphasis is placed primarily on avoidance of loss by production of forms that do not fail in use and are most closely suited to each task. These design directions would result in specialized and standardized tool forms.

For resource diversification,— I have suggested that the range of facilities would be diverse, but in ways different from the products of specialization. Diversity of facilities would result from construction of different forms to address the range of kinds of plants and animals exploited. Tool assemblages should not be diverse. Tools should be designed for multiple use and portability in order to be available when and where needed, and to be responsive to a variety of needs. This strategy reduces the probability of loss not by maximizing the reliability and efficiency of a tool but by creating a form that can deal with a range of potential uses and uncertainties.

As I have discussed above, all of these design options have advantages, but they also have disadvantages. In addition, "design option" is not synonymous with "tool type." Tools and weapons are combinations of greater and lesser emphasis on design options. There are always tradeoffs. If we can clearly separate the concept of design option from that of tool type, it should be possible to apply design options to developing expectations about suitable forms in different social, economic, and environmental contexts.

The design of tools and facilities is responsive to a number of different conditions, among which stress is but one. The specific conditions within which tools and facilities are being made, used, reused, and discarded must be considered for application of any of the ideas presented here.

No data set can be used alone as an infallible indicator of past behavior. Multiple lines of evidence are important in understanding the past. The design of tools and facilities is one of these sets of information. The inferences about strategies appropriate to ameliorating stress need further evaluation through the use of ethnobotanical and faunal data. Further, I have discussed little of the architectural and settlement implications of different strategies.

Given these limitations, the approach I have suggested may be useful for generating new ideas, and for testing old ideas with the data set that includes tools, weapons, and facilities.The concepts presented for describing technological strategies and design options can be extended to a variety of issues addressing how people use and manipulate their environment.

ACKNOWLEDGMENTS

This chapter was initially written for presentation at the Santa Fe Institute. I am grateful to the Institute and to the USDA Forest Service for sponsoring the workshop and my participation. Unpublished projectile point data were provided by Linda S. Cordell, Stephen H. Lekson, and Katherine Spielmann. These three and Cameron clarified specifics regarding the point collections.

Douglas Bamforth and Peter Bleed provided me with their manuscript in progress, as did Lisa Young. Brian Hayden and Andrew Christenson offered critical comments on an earlier draft. A brief version of one portion of this chapter was presented at the SAA annual meeting in 1992 at the invitation of Philip Carr. I thank all of these people for their data and innovative ideas.

REFERENCES

Akins, N. J.

1982 Temporal Variation in Faunal Assemblages from Chaco Canyon. In *Recent Research on Chaco Prehistory*, edited by W. J. Judge and J. D. Schelberg, pp. 225–240. Reports of the Chaco Center 8. Division of Cultural Research, USDI National Park Service.

Anell, Bengt

1969 Running Down and Driving of Game in North America. *Studia Ethnographica Usaliensia XXX*. Lund, Sweden.

Arnold, Dean E.

1985 *Ceramic Theory and Cultural Process.* Cambridge University Press, Cambridge.

Baksh, Michael, and Allen Johnson

1990 Insurance Policies Among the Machiguenga: An Ethnographic Analysis of Risk Management in a Non-Western society. In *Risk and Uncertainty in Tribal and Peasant Economies*, edited by Elizabeth Cashdan, pp. 193–227. Westview Press, Boulder.

Bamforth, Douglas B., and Peter Bleed
1991 The Concept of Risk and the Organization of Flaked Stone Technology. Ms. in possession of the author.

Berry, Michael S.
1982 *Time, Space, and Transition in Anasazi Prehistory.* University of Utah Press, Salt Lake City.

Bleed, Peter
1986 The Optimal Design of Hunting Weapons: Maintainability or Reliability. *American Antiquity* 51:737–47.

Binford, Lewis R.
1977 Forty-Seven Trips. In *Stone Tools as Cultural Markers*, edited by R. V. S. Wright, pp. 24–36. Australian Institute of Aboriginal Studies, Canberra.
1979 Organization and Formation Processes: Looking at Curated Technologies. *Journal of Anthropological Research* 35:255–73.

Blevins, Byron B.
1974 An Analysis of Qualitative Projectile Point Attributes with Regard to Resource Depletion, Tijeras Pueblo LA-581. Ms. in possession of the author.

Boydston, Roger A.
1989 A Cost-Benefit Study of Functionally Similar Tools. In *Time, Energy and Stone Tools*, edited by Robin Torrence, pp. 67–77. Cambridge University Press, Cambridge.

Braun, David P.
1983 Pots as Tools. In *Archaeological Hammers and Theories*, edited by A. Keen and J. Moore, pp. 107–34. Academic Press, New York.

Braun, David P., and Stephen Plog
1982 Evolution of 'Tribal' Social Networks: Theory and Prehistoric North American Evidence. *American Antiquity* 47:504–525.

Browne, Jim
1940 Projectile Points. *American Antiquity* 5:209–213.

Cameron, Catherine M.
1987 Chipped Stone Tools and Cores: An Overview of the 1982–1983 Field Seasons. In *Prehistoric Stone Technology on Northern Black Mesa, Arizona*, edited by William J. Parry and Andrew L. Christenson. Occasional Paper No. 12. Center for Archaeological Investigations, Southern Illinois University, Carbondale.

Cashdan, Elizabeth A.
1980 Egalitarianism Among Hunters and Gatherers. *American Anthropologist* 82:116–120.
1983 Territoriality Among Human Foragers: Ecological Models and an Application to Four Bushman Groups. *Current Anthropology* 24:47–66.
1990 *Risk and Uncertainty in Tribal and Peasant Economies.* Westview Press, Boulder (edited).

Christenson, Andrew L.
1987 Projectile Points: Eight Millennia of Projectile Change on the Colorado Plateau. In *Prehistoric Stone Technology on Northern Black Mesa, Arizona*, edited by William J. Parry and Andrew L. Christenson, pp. 143–198. Occasional Paper No. 12. Center for Archaeological Investigations, Southern Illinois University, Carbondale.

Cordell, Linda S.
1975 Predicting Site Abandonment at Wetherill Mesa. *The Kiva* 40(3):189–203.
1984 *Prehistory of the Southwest.* Academic Press, Orlando.

Cordell, Linda S., and George J. Gumerman (editors)
1989 *Dynamics of Southwestern Prehistory.* Smithsonian Institution Press, Washington, D.C.

Ellis, Christopher J.
1994 Factors Governing the Use of Stone Projectile Tips: The Ethnographic Data. Paper presented at the Annual meeting of the Society for American Archaeology, Anaheim.

Fish, Suzanne K., Paul R. Fish, Charles Miksicek, and John Madsen
1985 A Preliminary Analysis of Hohokam Settlement and Agriculture in the northern Tucson basin. In *Proceedings of the 1983 Hohokam Symposium*, Part I, edited by Alfred E. Dittert, Jr. and Donald E. Dove. Occasional Paper No. 2, pp. 75–100. Phoenix Chapter, Arizona Archaeological Society, Phoenix.
1986 Prehistoric Agave Cultivation in Southern Arizona. *Desert Plants* 7:107–100.

Flenniken, Jeffrey J., and Anan W. Raymond
1986 Morphological Projectile Point Typology: Replication Experimentation and Technological Analysis. *American Antiquity* 51: 603–614.

Flenniken, Jeffrey J. and Alan L. Stanfill
1980 A Preliminary Technological Examination of 20 Archaeological Sites Located During the Cultural Resource Survey of the Whitehorse Ranch Public Land Exchange. *Contract Abstracts and CRM Archeology* 1: 23–30.

Ford, Richard I.
1981 Gardening and Farming Before A.D. 1000: Patterns of Prehistoric Cultivation North of Mexico. *Journal of Ethnobiology* 1(1):6–27.

Goodyear, Albert C.
1989 A Hypothesis for the Use of Cryptocrystalline Raw Materials Among Paleoindian Groups of North America. In *Eastern Paleoindian Lithic Resource Use*, edited by C. J. Ellis and J. C. Lothrop, pp. 1–9. Westview Press, Boulder.

Graves, Michael W., William A. Longacre, and Sally J. Holbrook
1982 Aggregation and Abandonment at Grasshopper Pueblo, Arizona. *Journal of Field Archaeology* 9:193–206.

Green, Judi
1985 The Light Duty Gran Quivira Lithic Assemblage. Ms. on file, University of Iowa.

Hames, Raymond
1990 Sharing among the Yanomamo: Part I, The Effects of Risk. In *Risk and Uncertainty in Tribal and Peasant Economies*, edited by Elizabeth Cashdan, pp. 89–105. Westview Press, Boulder.

Hard, Robert J.
1986 *Ecological Relationships Affecting the Rise of Farming Economies: A Test from the American Southwest.* Ph.D. dissertation, University of New Mexico, Albuquerque. University Microfilms, Ann Arbor.
1990 Agricultural Dependence in the Mountain Mogollon. In *Perspectives on Southwestern Prehistory*, edited by Paul E. Minnis and Charles L. Redman, pp. 135–149. Westview Press, Boulder.

Hegmon, Michelle
1989 Risk Reduction and Variation in Agricultural Economies: A Computer Simulation of Hopi Agriculture. *Research in Economic Anthropology* 11:89–121.
1995 Variability in Food Production, Strategies of Storage and Sharing, and the Pithouse-to-Pueblo Transition in the Northern Southwest. This volume.

Herrington, Selma Laverne
1979 *Settlement Patterns and Water Control Systems of the Mimbres Classic Phase, Grant County, New Mexico.* Ph.D. dissertation. University of Texas, Austin. University Microfilms, Ann Arbor.

Hill, James N.
1985 Style: A Conceptual Evolutionary Framework. In *Decoding Prehistoric Ceramics*, edited by Ben A. Nelson, pp. 362–388. Southern Illinois University Press, Carbondale.

Hill, Kim, and Kristen Hawkes
1983 Neotropical Hunting Among the Ache of Eastern Paraguay. In *Adaptive Responses of Native Amazonians*, edited by R. Hames and W. Vickers, pp. 139–188. Academic Press, New York.

Horsfall, Gayle
1987 A Design Theory Pespective on Variability in Grinding Stones. In *Lithic Studies Among the Highland Maya*, edited by B. Hayden, pp. 332–378. The University of Arizona Press, Tucson.

Hunter-Anderson, R. D.
1986 *Prehistoric Adaptation in the American Southwest.* Cambridge University Press, Cambridge.

Jelinek, Arthur J.
1967 *A Prehistoric Sequence in the Middle Pecos Valley, New Mexico.* Anthropological Papers, No. 31. Museum of Anthropology, University of Michigan, Ann Arbor.

Jochim, Michael A.
1989 Optimization and Stone Tool Studies: Problems and Potential. In *Time, Energy and Stone Tools*, edited by Robin Torrence, pp. 106–120. Cambridge University Press, Cambridge.

Kaplan, Hillard, Kim Hill, and A. Magdalena Hurtado
1990 Risk, Foraging and Food Sharing Among the Ache. In *Risk and Uncertainty in Tribal and Peasant Economies*, edited by Elizabeth Cashdan, pp. 107–143. Westview Press, Boulder.

Keeley, Lawrence H.
1982 Hafting and Retooling: Effects on the Archaeological Record. *American Antiquity* 47:798–809.

Kelly, Robert L.
1983 Hunter-Gatherer Mobility Strategies. *Journal of Anthropological Research* 39:277–306.
1988 The Three Sides of a Biface. *American Antiquity* 53:717–734.

Kidder, Alfred Vincent
1932 *The Artifacts of Pecos.* Yale University Press, New Haven.

Koldehoff, Brad
1987 The Cahokia Flake Tool Industry: Socioeconomic Implications for Late Prehistory in the Central Mississippi Valley. In *The Organization of Core Technology*, edited by J. K. Johnston and C. A. Morrow, pp. 151–185. Westview Press, Boulder.

Lancaster, James W.
1983 *An Analysis of Manos and Metates from the Mimbres Valley, New Mexico.* Unpublished Master's Thesis, Department of Anthropology, University of New Mexico, Albuquerque.

Lekson, Stephen H.
1985 Points, Knives and Drills of Chaco Canyon. Unpublished Ms. in possession of the author.

Leonard, Robert D.
1986 *Patterns of Anasazi Subsistence: Faunal Exploitation, Subsistence Diversification, and Site Function in Northeastern Arizona.* Ph.D. dissertation, University of Washington, Seattle. University Microfilms, Ann Arbor.

Low, Bobbi S.
1990 Human Responses to Environmental Extremeness and Uncertainty: A Cross-Cultural Perspective. In *Risk and Uncertainty in Tribal and Peasant Economies*, edited by Elizabeth Cashdan, pp. 229–256. Westview Press, Boulder.

Matson, R. G., and Brian Chisholm
1986 Basketmaker II Subsistence: Carbon Isotopes and Other Dietary Indicators from Cedar Mesa, Utah. Paper presented at the Third Anasazi Symposium, Monument Valley, Utah.

Mauldin, Raymond

1991 Agricultural Intensification in the Mogollon Highlands. In *Mogollon V*, edited by P. H. Beckett. COAS Publishing, Las Cruces, NM.

Meltzer, David J.

1991 Altithermal Archaeology and Paleoecology at Mustang Springs, on the Southern High Plains of Texas. *American Antiquity* 56:236–267.

Minnis, Paul E.

1985a *Social Adaptation to Food Stress: A Prehistoric Southwestern Example.* University of Chicago Press, Chicago.

1985b Domesticating People and Plants in the Greater Southwest. In *Prehistoric Food Production in North America*, edited by R. I. Ford, pp. 309–340. Anthropological Papers No. 75. Museum of Anthropology, University of Michigan, Ann Arbor.

1989 Prehistoric Diet in the Northern Southwest: Macroplant Remains from Four Corners Feces. *American Antiquity* 54:543–563.

Morrow, Carol A.

1987 Blades and Cobden Chert: A Technological Argument for Their Role as Markers of Regional Identification During the Hopewell Period in Illinois. In *The Organization of Core Technology*, edited by J. K. Johnson and C. A. Morrow, pp. 119–150. Westview Press, Boulder.

Myers, Andrew

1989 Reliable and Maintainable Technological Strategies in the Mesolithic of Mainland Britain. In *Time, Energy and Stone Tools*, edited by Robin Torrence, pp. 78–91. Cambridge University Press, Cambridge.

Nelson, Margaret C.

1981 *Chipped Stone Analysis in the Reconstruction of Prehistoric Subsistence Practices: An Example from Southwestern New Mexico.* Ph.D. dissertation, University of California at Santa Barbara. University Microfilms, Ann Arbor.

1986 Chipped Stone Analysis: Food Selection and Hunting Behavior. In *Short-Term Sedentism in the American Southwest, The Mimbres Valley Salado*, edited by Ben A. Nelson and Steven A. LeBlanc. University of New Mexico Press, Albuquerque.

1991 *The Study of Technological Organization. Archaeological Method and Theory 3(1)*, edited by Michael B. Schiffer, pp. 57–100. University of Arizona Press, Tucson.

Nelson, Margaret C., and Heidi Lippineier

1993 Grinding-Tool Design as Conditioned by Land-Use Pattern. *American Antiquity* 58:286–305

Parks, S.

1995 Getting to the Point: The Physics and Morphology of Projectile Points. Unpublished Ms. in possession of author.

Parry, W. J., and R. L. Kelly
1987 Expedient Core Technology and Sedentism. In *The Organization of Core Technology*, edited by J. K. Johnson and C. A. Morrow, pp. 285–305. Westview Press, Boulder.

Plog, S.
1978 Social Interaction and Stylistic Similarity: A Reanalysis. In *Advances in Archaeological Method and Theory*, edited by M. B. Schiffer, pp. 143–282. Academic Press, New York.

Rautman, A. E.
1993 Resource Variability, Risk, and the Structure of Social Networks: An Example from the Prehistoric Southwest. *American Antiquity* 58:403–424.

Rautman, A. E.
1995 Risk, Reciprocity, and the Operation of Social Networks. This volume.

Reid, J. J.
1978 Response to Stress at Grasshopper Pueblo, Arizona. In *Discovering Past Behavior: Experiments in the Archaeology of the American Southwest*, edited by P. Grebinger, pp. 195–213. Gordon and Breach, New York.

Schiffer, M. B., and J. M. Skibo
1987 Theory and Experiment in the Study of Technological Change. *Current Anthropology* 28:595–622.

Shott, M.
1986 Technological Organization and Settlement Mobility: An Ethnographic Examination. *Journal Anthropological Research* 42: 15–51.

Skibo, J., M. Schiffer, E. Blimman, and D. Wilson
1991 Beginning of Anasazi Pottery: A Technological and Experimental Study. Unpublished Ms., National Science Foundation Proposal in possession of the author.

Smith, E. A., and R. Boyd
1990 Risk and Reciprocity: Hunter-Gatherer Socioecology and the Problem of Collective Action. In *Risk and Uncertainty in Tribal and Peasant Economies*, edited by E. Cashdan, pp. 167–191. Westview Press, Boulder.

Spielmann, K. A.
1983 Late Prehistoric Exchange Between the Southwest and Southern Plains. *Plains Anthropologist* 102(1):257–272.

Spielmann, K. A.
1995 Hunting, Gathering, and Health in the Prehistoric Southwest. This volume.

Torrence, R.
1983 Time Budgeting and Hunter-Gatherer Technology. In *Hunter-Gatherer Economy in Prehistory: A European Perspective*, edited by G. Bailey, pp. 11–22. Cambridge University Press, Cambridge.
1989 Re-tooling: Towards a Behavioral Theory of Stone Tools. In *Time, Energy and Stone Tools*, edited by R. Torrence, 57–66. Cambridge University Press, Cambridge.

Vivian, R. Gwinn
1974 Conservation and Diversion: Water-Control Systems in the Anasazi Southwest. In *Irrigation's Impact on Society*, edited by by T. Downing and M. Gibson, 95–112. Anthropological Papers 25. University of Arizona, Tucson.

Watson, P. G.
1991 Early Plant Cultivation in the Southwest and in the Eastern Woodlands: Pattern and Process. In *Puebloan Past and Present, Papers in Honor of Stewart Peckham*, edited by M. S. Duran And D. T. Kirkpatrick. The Archaeological Society of New Mexico, Albuquerque.

Watt, K. E. F.
1972 Man's Efficient Rush Toward Deadly Dullness. *Natural History* February: 74–82.

Wiessner, P.
1982 Risk, Reciprocity and Social Influences on !Kung San Economics. In *Politics and History in Band Societies*, edited by E. R. Leacock and R. B. Lee. Cambridge University Press, Cambridge.
1983 Style and Social Information in Kalahari San Projectile Points. *American Antiquity* 48: 253–276.

Wilkie, D. S., and B. Curran
1991 Why do Mbuti Hunters Use Nets? Ungulate Hunting Efficiency of Archers and Net Hunters in the Ituri Rain Forrest. *American Anthropologist* 93:680–689.

Wills, W. H.
1988 Early Agriculture and Sedentism in the American Southwest: Evidence and Interpretations *Journal of World Prehistory* 2:445–488.

Wills, W. H., and B. Huckell
1994 Economic Implications of Changing Land-Use Patterns in the Late Archaic. In *Themes in Southwest Prehistory*, edited by G. J. Gumerman, pp. 33–52. School of American Research Press, Santa Fe.

Winterhalder, B.
1990 Open Field, Common Pot: Harvest Variability and Risk Avoidance in Agricultural and Foraging Societies. In *Risk and Uncertainty in Tribal and Peasant Economies*, edited by E. Cashdan. Westview Press, Boulder.

Wobst, H. M.
1977 Stylistic Behavior and Information Exchange. In *For the Director: Research Easays in Honor of James B. Griffin*, edited by C. E. Cleland. Anthropological Papers 6. Museum of Anthropology, University of Michigan, Ann Arbor.

Woodburn, J.
1982 Egalitarian Socieites. *Man* 17:431–451.

Young, L. C.
1992 Pits, Rooms, Baskets, Pots: Storage Among Southwestern Farmers. Paper presented at the Southwest Symposium, Tucson, Arizona.

Alan P. Sullivan III
Department of Anthropology, University of Cincinnati, Cincinnati, OH 45221–0380

Risk, Anthropogenic Environments, and Western Anasazi Subsistence

The American Southwest has long been used to test models about the relationship between environment and culture (Gumerman 1988). Historically, this relationship has turned on the questions of (i) how societies provisioned themselves and (ii) whether variation in subsistence economies could explain other differences among these societies as well (Jorgensen 1983). With a new vocabulary, archaeologists have rephrased these long-standing questions in terms of models that stipulate cultural responses to "risk" (e.g., Fagan 1991:265–268), often employing data extrapolated from contemporary ethnological and environmental settings (Rautman 1993). Further, many of our current conceptions of environment/culture interaction in the prehistoric Anasazi Southwest are based on premises regarding plant-use and human behavior (Wills 1992) that are rarely scrutinized critically. In this chapter, I examine the foundations of risk-based models, evaluate their consequences, and sketch an alternative model that explores the implications of systematic burning and wild-resource production for Western Anasazi subsistence organization.

RISK AND RESOURCES: DEEPLY ROOTED PREMISES

Since their initial scientific exploration, the Colorado Plateaus have been viewed as innately hostile to human livelihood (e.g., Powell 1875). Consequently, human societies that have resided there generally have been considered under stress (e.g., Minnis 1985) or, worse yet, at risk (*sensu* Torrence 1989:59; Winterhalder 1990:72). Interestingly, Lewis Henry Morgan captured the essence of this view more than 100 years ago: "The pueblos now in ruins throughout the original area of New Mexico, and for some distance north of it, testify to the perpetual struggle of the former to maintain their ground, as well as prove the insecurity in which they lived" (1965 [1881]:147).

People often inhabit environments where resources are spatially and temporally inconstant—in other words, risky (e.g., Dean 1988:29–30; Euler et al. 1979; Hevly 1988:96; Schlanger and Wilshusen 1993). It is generally presumed, furthermore, that wild-resource fluctuations adversely affect economic decision making because little control can be exercised over two variables (cf. Wills 1988:478–479): (i) availability (where a resource is likely to produce) and (ii) productivity (yield). Although they have their own set of problems (Bradfield 1971), agricultural systems allegedly were adopted because they allowed humans to establish some control over edible-resource availability and productivity (Minnis 1992:132). Despite some disagreement regarding the origins of agricultural systems on the Colorado Plateaus (Matson 1991; Wills 1992), archaeologists generally have thought that the Western Anasazi ultimately became dependent to varying degrees upon domesticated plant production and, consequently, developed a range of problem-solving strategies to grow cultigens in unpredictable or marginal environments (Bye and Shuster 1984; Dean et al. 1985:542–544). Many of the technological innovations that the Western Anasazi developed, such as terraces, alignments, and checkdams (Stewart and Donnelly 1943), it has been argued, were designed to "buffer" agricultural systems against threats to a successful harvest (Plog et al. 1988).

Risk-based models also imply that the Western Anasazi still were confronted with constant challenges to their livelihood and that, in order to secure a living in many areas of the Plateau Southwest, agricultural production would have been a necessity (Wills et al. 1994). However, wild resources were still an important component in the food supply of agriculturally dependent systems because of "their relative independence of factors responsible for agricultural loss" (O'Shea 1989:58; also Dean 1988:35; Powell 1983). Presumably, the evolutionary histories of wild plant species have adapted them to tolerate growing conditions that are unfavorable to cultivars. Such a tolerance would be particularly important when agriculturally dependent populations, like the Western Anasazi, became "stressed" during periods of (i) environmental variability adverse to cultivar production (Euler et al. 1979) or of (ii) increasing population pressure (sensu Keeley 1988; Braun and Plog 1982:513; Dean et al. 1994; Euler 1988). The clear implication, however, is that wild resources became of secondary importance once Western Anasazi populations had made a commitment to agricultural production.

One theme that penetrates risk-based models of Western Anasazi subsistence is that resource production occurs in environments that are either unmodified or minimally modified (e.g., Leonard 1989). In the remainder of this paper, I explore the theoretical consequences that environmental modification poses for risk-based models of Western Anasazi subsistence organization by investigating two factors: (i) the economic effects of burning the understory of pinyon-juniper woodland and (ii) the production of wild resources.

PRODUCTION, ANTHROPOGENIC ENVIRONMENTS, AND RISK

As Roy Ellen (1982) reminds us, it is production, executed through particular combinations of subsistence methods (Layton et al. 1991), that integrates ecology and economy. Among the many types of landscape modifications that are created by people (Thomas 1956), none has been more prevalent or more potentially productive than the anthropogenic environments that emerge from systematic burning (Stewart 1956). The ecological effects of the application of fire—specifically, low-intensity understory burns that do not consume the canopy—to various kinds of natural environments are well known (Kozlowski and Ahlgren 1974; Wright and Bailey 1982). In addition to many other benefits, the principal economic reason that burning has been so pervasive worldwide is that the productivity of biotic communities is enhanced enormously following the application of fire (Kohler 1992a:239; Kramp et al. 1983). It is no surprise, therefore, that the human use of fire has great time depth (James 1989) and is geographically widespread (e.g., Barrett 1980; Gould 1971; Lewis 1972, 1982; Mellars 1976; Patterson and Sassaman 1988; Smith 1988:152–153).[1]

FIRE IN A PINYON-JUNIPER WOODLAND: ECOLOGY AND ECONOMY

Pinyon-juniper woodland is a widespread vegetation community (Barger and Ffolliett 1972) that, during prehistoric times, sustained the bulk of Western Anasazi occupation on the Colorado Plateaus (Euler et al. 1979:1095; Lanner 1981:66; Tainter 1984). The burning of a pinyon-juniper woodland understory produces two predictable effects (Table 1; West and Van Pelt 1987) that would have been

[1] For monograph-length case studies that explore the subsistence implications of systematic burning, see Lewis (1973) for California, Dobyns (1981) for the Southwest, Pyne (1982) for North America, and Gouldsblom (1992) for Europe.

TABLE 1 Time-line of successional stages in a burned pinyon-juniper woodland.[1]

Stage:	Annual	Perennial Grass/Forb	Shrub	Shrub/ Open Tree	Climax P-J Forest
Years:	2	4	25	100	300

[1] After Erdman (1970:18), Barney and Frischknecht (1974:95), Everett and Ward (1984:60), and Everett (1987).

particularly beneficial for Western Anasazi subsistence (Bye 1981:111–113; Martin et al. 1991:65): (i) the diversity of edible plant species, both perennials and annuals (Everett and Ward 1984:62), expands and (ii) biomass increases (Despain 1987). West (1984:1308–1309) notes that, with respect to annuals in particular, it is not uncommon for there to be an increase of upwards of 20% per hectare after burning. Controlled studies (e.g., Barney and Frischknecht 1974:94; Wright et al. 1979:29) show that edible annual forbs such as tansy-mustard (*Descurainia sophia* [Bohrer 1973:435]), sunflower (*Helianthus* sp. [Adams 1980:32–33]), and goosefoot (*Chenopodium* sp. [Minnis 1984:7])[2] all appear within one year after a fire. Similarly, edible perennial herbs, such as Indian ricegrass (*Oryzopsis hymenoides* [Bohrer 1975]), *Agropyron* sp. (Doebley 1984:54), and bluegrass (*Poa* sp. [Doebley 1984:57]), emerge early after a fire and come to dominate a burned area within the first five or six years (Barney and Frischknecht 1974:96; Erdman 1970:19).[3] Interestingly, Indian ricegrass begins growing within only three weeks after burning (Wright et al. 1979:26).

RISK IN ANTHROPIC HABITATS

If it can be assumed that, in comparison to their modern structure and distribution (Ronco 1987), "pre-Columbian woodlands were likely more open and savanna-like largely because of fairly frequent fires, some of which were due to Amerinds" (West 1984:1310), then many of the supposed subsistence-related consequences of risk, enumerated above, are simply theoretical chimeras (Bailey 1981:6–7). For example, because people control where anthropogenic environments are established, the spatial component of risk—availability—is reduced considerably. Also, because they

[2] Wright et al. (1979:29) note that goosefoot (*Chenopodium* sp.) experiences "enormous relative increases" within the first weeks after a fire.

[3] Another important post-burn perennial is dropseed (*Sporobolus* sp. [Wright et al. 1979:14; Doebley 1984:57]).

can control the extent of burning, people can influence productivity (yield). In conjunction with the seasonal distribution of important edible annuals, such as *Lipidium* (winter), *Amaranthus* (spring-summer), and *Portulaca* (summer-fall), fire management virtually could ensure no gaps in the food supply (see also Bohrer 1975; Doebley 1984) or sharply reduce the possibility of resource deficits (Martin 1994:91). Unless these practices were managed prudently, obviously, the sustainability of such a land-use pattern ultimately would have been compromised by excessive population growth, the emergence of territorial boundaries, or habitat degradation (cf. Kohler 1992b).

INTERPRETING THE WESTERN ANASAZI ARCHAEOECONOMIC RECORD

PROBLEMATIC PRODUCTIVITY ESTIMATES

Archaeologists may have exaggerated the magnitude of any inherent risk in Western Anasazi economies because risk-based models have chronically and systematically underestimated wild-resource productivity in two ways (e.g., Martin et al. 1991:66). First, approximations of wild-resource productivity today (e.g., Ford 1984; Wills 1992) often are derived from historically modified (Gifford 1987) or degraded environments (Cooper 1960; West 1984) and from biotic communities with diminished species diversity (Bohrer 1975:206; Bye 1985:379). Second, the drastic modification of traditional plant-use patterns during post-contact times (e.g., Fogg 1965:103) affects the analogical usefulness of risk-based economic models that incorporate ethnographic data (e.g., Hegmon 1989a). The concatenation of these potential biases has skewed our views of Western Anasazi economic behavior, especially with respect to assertions regarding the lack of human control over the availability and productivity of wild plants grown in anthropogenic environments (Kehoe 1981). Hence, some of the proposed solutions to presumed subsistence risk, such as elaborate forms of social interaction (Hegmon 1989b) and mobility (Rautman 1993), may be epiphenomenal. In contrast, a number of studies suggests rather strongly that variation in arboreal pollen/nonarboreal pollen ratios (Edwards 1993), soil texture and chemistry (Courty et al. 1989:107–111,129), relative ubiquity of "fuel" taxa (Kohler and Matthews 1988), and thermoluminescence values (Rowlett 1991a), is ultimately attributable to local occupation and abandonment patterns that affected vegetation community composition (also Cummings 1994:135; Floyd and Kohler 1990). Burning would have been a major contributing cause of such variation in these phenomena (Hevly 1988:103; Matson et al. 1988:258; Rowlett 1991b).

PRODUCTION AND CONSUMPTION

Erratic attention regarding how the archaeoeconomic record arises (Minnis 1981; Pearsall 1988), furthermore, may be responsible for one of the major interpretive problems affecting studies of Western Anasazi subsistence—data that conventionally have been interpreted as evidence for production are, in fact, indicative of consumption. This is a problem because the relationship between production and consumption is asymmetric (cf. Fish and Donaldson 1991). Production strategies cannot be reliably inferred from samples that originate in consumption contexts because the processes that affect the composition of consumption assemblages, such as occupation mode and abandonment mode (Cameron and Tomka 1993), may be independent of contexts of production (Hastorf 1988).

WILD-RESOURCE PRODUCTION IN THE UPPER BASIN

To illustrate these points, I present some evidence for Western Anasazi wild-resource production. This evidence comes from recent excavations at two nonarchitectural sites—MU 235 and MU 236—located in the Upper Basin, which is a downfaulted block of the Coconino Plateau along the eastern south rim of the Grand Canyon (Figure 1). At these sites, the most obvious features, which heretofore have not been excavated, are piles of fire-cracked rock (Figure 2). Radiocarbon dates and associated ceramics imply that these features are prehistoric in age and are affiliated with the Western Anasazi (Sullivan 1992:209–213). Contrary to expectations, excavation of three piles failed to reveal any evidence of roasting pits under or adjacent to them (cf. Euler 1967); the piles rest directly on a prehistoric occupation surface. Charred seeds of Indian ricegrass (*Oryzopsis hymenoides*), buckwheat (*Eriogonum*), *cheno-ams*, and purslane (*Portulaca*) were found in flotation samples from the fire-cracked-rock piles (Sullivan 1992:215). As noted above, these wild plants were used extensively both prehistorically and historically on the Colorado Plateaus (e.g., Doebley 1984; Hutira 1986), and in the Great Basin (Winter and Hogan 1986). In addition, Site MU 235 disclosed *in situ* artifact concentrations that were 5–10 cm. below the present ground surface (Figure 3). These concentrations consist of patterned arrangements of complete manos, mano fragments, metates, metate fragments, cores, retouched pieces (all with clear signs of wear), and Tusayan Grayware jar fragments. Additional processing features and artifacts were discovered in situ at Site MU 236, as well (Figure 4). The form of the fire-cracked-rock piles, their paleobotanical contents, the patterned artifact arrays, as well as the disarticulated condition of the fire-cracked rock itself (Latas 1992:213), suggests that they are remains of edible-resource production locations, perhaps used repeatedly.[4]

[4] See Sullivan (1992:214–216) for details regarding the processing activities themselves.

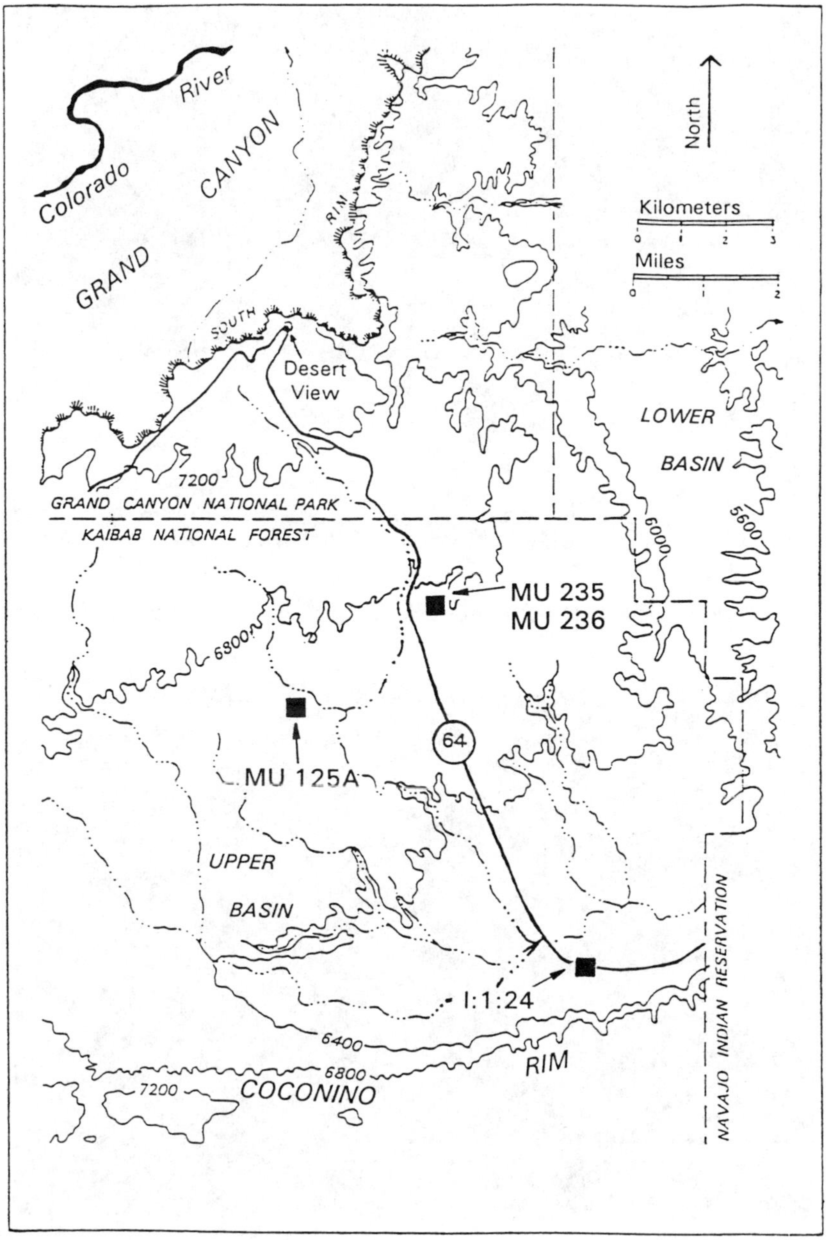

FIGURE 1 Map of the Upper Basin showing locations of sites MU 125A, MU 235, MU 236, and Site 24 (AZ I:1:24 [ASM]).

FIGURE 2 Oblique view of fire-cracked-rock pile (ca. 6m. × 4m.) at site MU 236 prior to excavation.

FIGURE 3 In situ groundstone artifacts at site MU 235 resting on the exposed prehistoric occupation surface. The grinding surface of the slab metate (top center) is in contact with the occupation surface; white flagging-tape marks manos or mano fragments.

FIGURE 4 Processing feature at site MU 236 before excavation; object at lower right is a metate fragment in contact with the prehistoric occupation surface.

FIGURE 5 Isolated cluster of three slab metates found during intensive survey of the Upper Basin.

Groundstone assemblages from Sites MU 235 and 236 contain neither a trough metate nor an identifiable trough-metate fragment—the form of metate conventionally associated with maize processing in the Southwest (Morris 1990). In addition, our intensive survey of 12 km.^{2} of the surrounding countryside has found predominantly slab or basin metates (complete and fragmentary), such as the isolated cluster of three slab metates seen in Figure 5. In the Upper Basin, trough metates and trough metate fragments have been found exclusively at architectural sites (Becher 1992; Sullivan 1986). Interestingly, in her analysis of groundstone artifacts from habitation sites in southwestern Colorado, Schlanger (1991:463) remarked that "slab and basin metates are relatively rare in the Dolores collection." These patterned differences in metate form support the hypothesis that the role of wild resources in Western Anasazi subsistence economies has been underestimated, for two reasons (cf. Powell 1988:179). First, our economic models have incorporated information that is skewed in favor of consumption rather than production locales in Western Anasazi landscapes (cf. Floyd and Kohler 1990:154). And second, subsistence inferences have been based on consumption-oriented rather than production-oriented assemblages.

The palynological data presented in Figure 6 support these arguments. Production is represented by the pollen percentages from Sites 24 (Homburg 1992) and 125A (Sullivan 1995), which are two rock-alignment complexes in the Upper Basin, and from the Merriam Crater (Berlin et al. 1977) and Sunset Crater (Berlin et al. 1990; Schaber and Gumerman 1969) ash-ridge fields located about 50 miles southeast of the Upper Basin. Consumption is represented by the pollen percentages from the well-preserved colon contents of a Kayenta Anasazi burial near Black Mesa, which is about 90 miles northwest of the Upper Basin (Reinhard et al. 1992). As this figure shows, production and consumption of corn may have been neither as common nor as widespread as once thought. The common denominator in the Upper Basin (Sites 24 and 125A) and Flagstaff (Merriam Crater and Sunset Crater) samples is the production of wild plants—notice the conspicuously low occurrence of corn pollen (Zea). It is not unreasonable to suggest, therefore, that variation in archaeobotanical (Gasser 1982; Toll 1988) and archaeoscatological (Minnis 1989) assemblages from architectural sites is *consumption-dependent* rather than *production-dependent* (Hawkins 1992:67). This may be an important distinction because, in comparison to production assemblages, consumption assemblages are highly likely to have been affected by all sorts of post-production processes such as draw-down or depletion, seasonal food preferences, variation in occupation mode, variation in abandonment mode (Sullivan 1987), and, of course, distribution and exchange—factors that would have little or no effect on the composition of production assemblages. Indeed, all of these data are consistent with the proposition that, until recently, archaeologists have downplayed the magnitude of wild-resource production in Western Anasazi subsistence economies. It is little wonder, then, that small-scale, economically autonomous populations on the Colorado Plateaus have been viewed as "at risk."

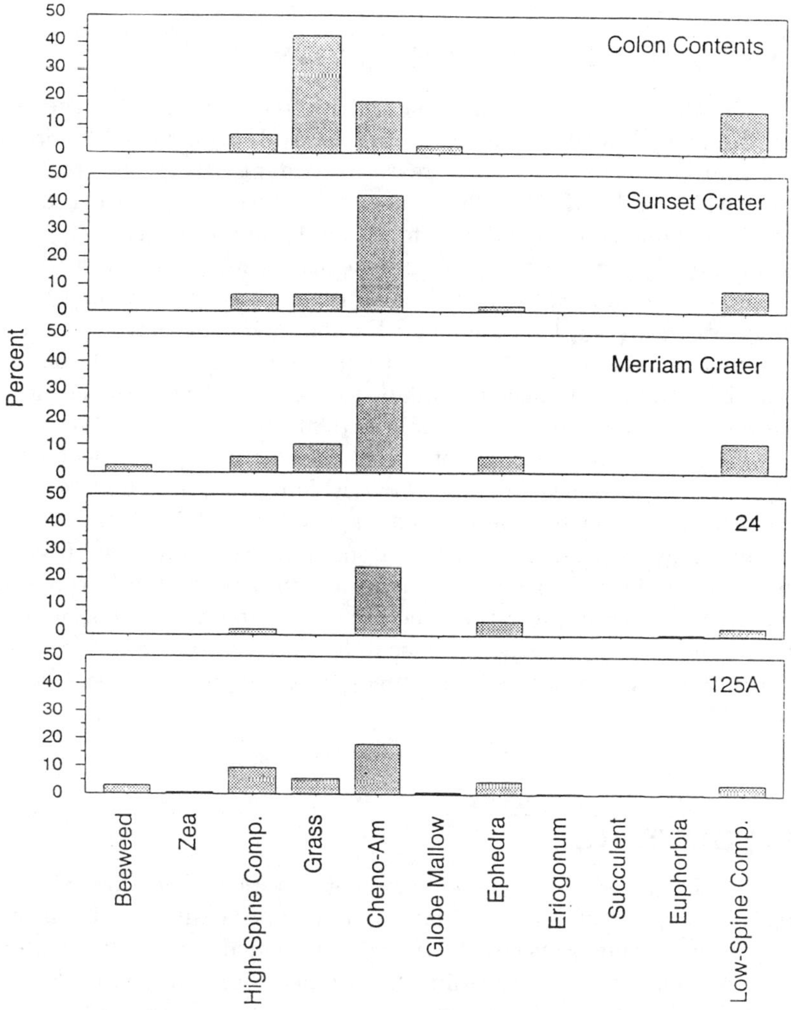

FIGURE 6 Bar charts of relative frequencies of economic pollen species from the colon remains of a Pueblo III burial in the Klethla Valley, ash-ridge fields near Sunset Crater and Merriam Crater east of Flagstaff, and two rock-alignment sites (24 and 125A) in the Upper Basin. Note the near absence of maize (Zea) in these archaeopalynological data.

RISK AND SOUTHWESTERN PREHISTORY

The importance of risk in decision making ought to be reconsidered with the following factors in mind. Models that focus on economic production should incorporate the likelihood that prehistoric societies create and adapt to anthropogenic environments (Ellen 1982:14–15; Kohler 1992a) whose successional stages can be managed as integral components of the annual food-supply system (Crites 1987). That Western Anasazi populations probably exercised considerable control over their environments through the judicious application of fire is ethnologically plausible and archaeologically testable (e.g., Caseldine and Hatton 1993). In fact, it would have been irrational not to burn systematically and repeatedly, knowing that concentrations of plants with edible seeds, such as Indian ricegrass, sunflower, and cheno-ams, were sure to follow the smoke and flames (Bohrer 1983). In view of how Western Europeans and Americans have conceived of fire as an inherently malevolent force to be controlled or avoided whatever the cost (Gouldsblom 1992), it is understandable that archaeologists might have some difficulty appreciating "the degree to which Indian economies were dependent on fire" (Pyne 1982:71). However, if Western Anasazi populations indeed managed their habitats with controlled burning, then many puzzles of Southwestern prehistory, such as punctuated regional occupational sequences, for example, may have their origins in fruitful economic processes, rather than in the presumed risks associated with precarious carrying capacities.[5]

ACKNOWLEDGMENTS

I thank Jeffrey S. Dean for his extended comments on an earlier draft of this paper that caused me to rethink some of my views on the feasibility of burning in a pinyon-juniper woodland. Rebecca A. Hawkins provided a much-needed external perspective on general issues regarding subsistence theory in archaeology. The manuscript also benefited considerably from the editorial comments of Joseph A. Tainter.

[5] Compare the paper by Jeffrey Dean, this volume.

REFERENCES

Adams, K. R.
1980 *Pollen, Parched Seeds and Prehistory: A Pilot Investigation of Prehistoric Plant Remains from Salmon Ruin, a Chacoan Pueblo in Northwestern New Mexico.* Contributions in Anthropology, Vol. 9. Eastern New Mexico University, Portales.

Bailey, G. N.
1981 Concepts of Resource Exploitation: Continuity and Discontinuity in Palaeoeconomy. *World Archaeology* 13:1–15.

Barger, R. L. and P. F. Folliott
1972 Physical Characteristics and Utilization of Major Woodland Tree Species in Arizona. Research Paper RM-83. USDA Forest Service, Rocky Mountain Forest and Range Experiment Station.

Barney, M. A. and N. C. Frischknecht
1974 Vegetation Changes Following Fire in the Pinyon-Juniper Type of West-Central Utah. *Journal of Range Management* 27:91–96.

Barrett, S. W.
1980 Indians & Fire. *Western Woodlands* 6:17–21.

Becher, M. E.
1992 Archaeological Investigations at a Pueblo II Kayenta Anasazi Ruin Located in the Tusayan Ranger District, Kaibab National Forest, Arizona. Ms. on file, Kaibab National Forest, Williams, AZ.

Berlin, G. L., J. R., Ambler, R. H. Hevly, and G. G. Schaber
1977 Identification of a Sinagua Agricultural Field by Aerial Thermography, Soil Chemistry, Pollen/Plant Analysis, and Archaeology. *American Antiquity* 42:588–600.

Berlin, G. L., D. E. Salas, and P. R. Geib
1990 A Prehistoric Sinagua Agricultural Site in the Ashfall Zone of Sunset Crater, Arizona. *Journal of Field Archaeology* 17:1–16.

Bohrer, V. L.
1973 Ethnobotany of Point of Pines Ruin, Arizona W:10:50. *Economic Botany* 27:423–437.
1975 The Prehistoric and Historic Role of the Cool-Season Grasses in the Southwest. *Economic Botany* 29:199–207.
1983 New Life from Ashes: The Tale of the Burnt Bush (*Rhus trilobata*). *Desert Plants* 5:122–125.

Bradfield, M.
1971 *The Changing Pattern of Hopi Agriculture.* Occasional Paper No. 30. Royal Anthropological Institute of Great Britain and Ireland.

Braun, D. P. and S. Plog
1982 Evolution of 'Tribal' Social Networks: Theory and Prehistoric North American Evidence. *American Antiquity* 47:504–525.

Bye, R. A., Jr.

1981 Quelites–Ethnoecology of Edible Plants–Past, Present, and Future. *Journal of Ethnobiology* 1:109-123.

1985 Botanical Perspectives of Ethnobotany of the Greater Southwest. *Economic Botany* 39:375–386.

Bye, R. A., Jr. and R. Shuster

1984 Developing an Integrated Model for Contemporary and Archaeological Agricultural Subsistence Systems. In *Prehistoric Agricultural Strategies in the American Southwest*, edited by S. K. Fish and P. R. Fish, pp. 125–145. Anthropological Research Papers No. 33. Arizona State University, Tucson.

Cameron, C. M. and S. A. Tomka (eds.)

1993 *Abandonment of Settlements and Regions: Ethnoarchaeological and Archaeological Approaches.* Cambridge University Press, Cambridge.

Caseldine, C. and J. Hatton

1993 The Development of High Moorland on Dartmoor: Fire and the Influence of Mesolithic Activity on Vegetation Change. In *Climate Change and Human Impact on the Landscape*, edited by F. M. Chambers, pp. 119–131. Chapman and Hall, London.

Cooper, C. F.

1960 Changes in Vegetation, Structure, and Growth of Southwestern Pine Forests Since White Settlement. *Ecological Monographs* 30:129–164.

Courty, M. A., P. Goldberg, and R. MacPhail

1989 *Soils and Micromorphology in Archaeology.* Cambridge University Press, Cambridge.

Crites, G. D.

1987 Human-Plant Mutualism and Niche Expression in the Paleoethnobotanical Record: A Middle Woodland Example. *American Antiquity* 52:725–740.

Cummings, L. S.

1994 Anasazi Diet: Variety in the Hoy House and Lion House Coprolite Record and Nutritional analysis. In *Paleonutrition: The Diet and Health of Prehistoric Americans*, edited by K. D. Sobolik, pp. 134–150. Southern Illinois University Press, Carbondale.

Dean, J. S.

1988 A Model of Anasazi Behavioral Adaptation. In *The Anasazi in a Changing Environment*, edited by G. J. Gumerman, pp. 25–44. Cambridge University Press, Cambridge.

Dean, J. S., R. C. Euler, G. J. Gumerman, F. Plog, R. H. Hevly, and T. N. V. Karlstrom

1985 Human Behavior, Demography, and Paleoenvironment on the Colorado Plateaus. *American Antiquity* 50:537–554.

Dean, J. S., W. H. Doelle, and J. D. Orcutt
1994 Adaptive Stress, Environment, and Demography. In *Themes in Southwest Prehistory*, edited by G. J. Gumerman, pp. 53–86. School of American Research Press, Santa Fe.
Despain, D. W.
1987 History and Results of Prescribed Burning of Pinyon-Juniper Woodland on the Hualapai Reservation in Arizona. In *Proceedings—Pinyon-Juniper Conference*, compiled by R. L. Everett, pp. 145–151. General Technical Report INT-215. USDA Forest Service, Intermountain Research Station.
Doebley, J. F.
1984 Seeds of Wild Grasses: A Major Food of Southwestern Indians. *Economic Botany* 38:52–64.
Dobyns, H. F.
1981 *From Fire to Flood: Historic Human Disturbance of Sonoran Desert Riverine Oases.* Anthropological Papers No. 20. Ballena Press, Socorro, NM.
Edwards, K. J.
1993 Models for Mid-Holocene Forest Farming for North-west Europe. In *Climate Change and Human Impact on the Landscape*, edited by F. M. Chambers, pp. 133–145. Chapman and Hall, London.
Ellen, R.
1982 *Environment, Subsistence and System: The Ecology of Small-Scale Social Formations.* Cambridge University Press, Cambridge.
Erdman, J. A.
1970 *Pinyon-Juniper Succession after Natural Fires on Residual Soils of Mesa Verde, Colorado.* Science Bulletin, Biological Series XI (2). Brigham Young University, Provo, UT.
Euler, R. C.
1967 The Canyon Dwellers. *The American West* 4:22–27;67–71.
1988 Demography and Cultural Dynamics on the Colorado Plateaus. In *The Anasazi in a Changing Environment*, edited by G. J. Gumerman, pp. 192–229. Cambridge University Press, Cambridge.
Euler, R. C., G. J. Gumerman, T. N. V. Karlstrom, J. S. Dean, and R. H. Hevly
1979 The Colorado Plateaus: Cultural Dynamics and Paleoenvironment. *Science* 205:1089–1101.
Everett, R. L.
1987 Plant Response to Fire in the Pinyon-Juniper Zone. In *Proceedings—Pinyon-Juniper Conference*, compiled by R. L. Everett, pp. 152–157. General Technical Report INT-215. USDA Forest Service, Intermountain Research Station.
Everett, R. L. and K. Ward
1984 Early Plant Succession on Pinyon-Juniper Controlled Burns. *Northwest Science* 58:57–68.

Fagan, B. M.
1991 *Ancient North America: The Archaeology of a Continent.* Thames and Hudson, London.

Fish, S. K. and M. Donaldson
1991 Production and Consumption in the Archaeological Record: A Hohokam Example. *Kiva* 56:255–275.

Floyd, M. L. and T. A. Kohler
1990 Current Productivity and Prehistoric Use of Pinon (*Pinus edulis*, Pinaceae) in the Dolores Archaeological Project Area, Southwestern Colorado. *Economic Botany* 44:141–156.

Fogg, G. G.
1965 The Pinyon Pines and Man. *Economic Botany* 20:103–105.

Ford, R. I.
1984 Ecological Consequences of Early Agriculture in the Southwest. In *Papers on the Archaeology of Black Mesa, Arizona, Volume II*, edited by S. Plog and S. Powell, pp. 127–138. Southern Illinois University Press, Carbondale.

Gasser, R. E.
1982 Anasazi Diet. In *The Coronado Project Archaeological Investigations, The Specialists' Volume: Biocultural Analyses*, compiled by R. E. Gasser, pp. 8–95. Research Paper No. 23. Museum of Northern Arizona, Tucson.

Gifford, G. P.
1987 Myths and Fables and the Pinyon-Juniper Type. In *Proceedings—Pinyon-Juniper Conference*, compiled by R. L. Everett, pp. 34–37. General Technical Report INT-215. USDA Forest Service, Intermountain Research Station.

Gould, R. A.
1971 Uses and Effects of Fire Among the Western Desert Aborigines of Australia. *Mankind* 8:14–24.

Gouldsblom, J.
1992 *Fire and Civilization.* Penguin Books, London.

Gumerman, G. J.
1988 A Historical Perspective on Environment and Culture in Anasazi Country. In *The Anasazi in A Changing Environment*, edited by G. J. Gumerman, pp. 1–24. Cambridge University Press, Cambridge.

Hastorf, C. A.
1988 The Use of Paleoethnobotanical Data in Prehistoric Studies of Crop Production, Processing, and Consumption. In *Current Paleoethnobotany: Analytical Methods and Cultural Interpretations of Archaeological Plant Remains*, edited by C. A. Hastorf and V. S. Popper, pp. 119–144. University of Chicago Press, Chicago.

Hawkins, R. A.
1992 Subsistence Inferences from Woodland and Mississippian Ceramics: The Central Ohio Valley, Circa 1000 B.C.–A.D. 1200. In *Long-Term Subsistence Change in Prehistoric North America*, edited by D. R. Croes, R. A. Hawkins, and B. L. Isaac, pp. 47–76. JAI Press, Greenwich, CT.

Hegmon, M.
1989a Risk Reduction and Variation in Agricultural Economies: A Computer Simulation of Hopi Agriculture. In *Research in Economic Anthropology* Vol. 11, edited by B. L. Isaac, pp. 89–121. JAI Press, Greenwich, CT.
1989b The Styles of Integration: Ceramic Style and Pueblo I Integrative Architecture in Southwestern Colorado. In *The Architecture of Social Integration in Prehistoric Pueblos*, edited by W. D. Lipe and M. Hegmon, pp. 12-5-141. Occasional Papers No. 1. Crow Canyon Archaeological Center.

Hevly, R. H.
1988 Prehistoric Vegetation and Paleoclimates on the Colorado Plateaus. In *The Anasazi in a Changing Environment*, edited by G. J. Gumerman, pp. 92–118. Cambridge University Press, Cambridge.

Homburg, J. A.
1992 Soil Fertility Study. In *Investigations at Lee Canyon: Kayenta Anasazi Farmsteads in the Upper Basin, Coconino County, Arizona*, edited by S. M. Whittlesey and J. A. Homburg, pp. 145–161. Technical Series No. 38. Statistical Research, Inc.

Hutira, J.
1986 Analysis of Plants, Pollen, and Coprolites. In *A Cross Section of Grand Canyon Archaeology: Excavations at Five Sites Along the Colorado River*, edited by A. T. Jones, pp. 269–322. Publications in Anthropology No. 28. USDI National Park Service, Western Archeological and Conservation Center.

James, S. R.
1989 Hominid Use of Fire in the Lower and Middle Pleistocene. *Current Anthropology* 30:1–26.

Jorgensen, J. G.
1983 Comparative Traditional Economies and Ecological Adaptations. In *Handbook of North American Indians, Volume 10, Southwest*, edited by A. Ortiz, pp. 684–710. Smithsonian Institution, Washington, D.C.

Keeley, L. H.
1988 Hunter-Gatherer Economic Complexity and 'PopulationPressure': A Cross-Cultural Analysis. *Journal of Anthropological Archaeology* 7:373–411.

Kehoe, A. B.
1981 Revisionist Anthropology: Aboriginal North America. *Current Anthropology* 22:503–517.

Kohler, T. A.

1992a The Prehistory of Sustainability. *Population and Environment: A Journal of Interdisciplinary Studies* 13:237–242.

1992b Prehistoric Human Impact on the Environment in the Upland North American Southwest. *Population and Environment: A Journal of Interdisciplinary Studies* 13:255–268.

Kohler, T. A. and M. H. Matthews

1988 Long-Term Anasazi Land Use and Forest Reduction: A Case Study from Southwest Colorado. *American Antiquity* 53:537–564.

Kozlowski, T. T. and C. E. Ahlgren

1974 *Fire and Ecosystems.* Academic Press, New York.

Kramp, B. A., D. R. Patton, and W. W. Brady

1983 *The Effects of Fire on Wildlife Habitat and Species.* USDA Forest Service, Southwestern Region, Albuquerque.

Lanner, R. M.

1981 *The Pinon Pine.* University of Nevada Press, Reno.

Latas, T. W.

1992 An Analysis of Fire-Cracked Rock: A Sedimentological Approach. In *Deciphering a Shell Midden*, edited by J. K. Stein, pp. 211–237. Academic Press, New York.

Layton, R., R. Foley, and E. Williams

1991 The Transition Between Hunting, Gathering, and the Specialized Husbandry of Resources. *Current Anthropology* 32:255–269.

Leonard, R. D.

1989 Resource Specialization, Population Growth, and Agricultural Production in the American Southwest. *American Antiquity* 54:491–503.

Lewis, H. T.

1972 The Role of Fire in the Domestication of Plants and Animals of Southwest Asia: A Hypothesis. *Man* 7:195–222.

1973 *Patterns of Indian Burning in California: Ecology and Ethnohistory.* Anthropological Papers No. 1. Ballena Press, Ramona, CA.

1982 Fire Technology and Resource Management in Aboriginal North America and Australia. In *Resource Managers: North American and Australian Hunter-Gatherers*, edited by N. M. Williams and E. S. Hunn, pp. 45–68. American Association for the Advancement of Science, Washington, D.C.

Martin, D. L.

1994 Patterns of Diet and Disease: Health Profiles for the Prehistoric Southwest. In *Themes in Southwest Prehistory*, edited by G. J. Gumerman, pp. 87–108. School of American Research Press, Santa Fe.

Martin, D. L., A. H. Goodman, G. J. Armelagos, and A. L. Magennis

1991 *Black Mesa Anasazi Health: Reconstructing Life from Patterns of Death and Disease.* Occasional Paper No. 14. Southern Illinois University at Carbondale.

Matson, R. G.
1991 *The Origins of Southwestern Agriculture.* University of Arizona Press, Tucson.
Matson, R. G., W. D. Lipe, and W. R. Haase IV
1988 Adaptational Continuities and Occupational Discontinuities: The Cedar Mesa Anasazi. *Journal of Field Archaeology* 15:245–264.
Mellars, P.
1976 Fire Ecology, Animal Populations and Man: A Study of Some Ecological Relationships in Prehistory. *Proceedings of the Prehistoric Society* 42:15–45.
Minnis, P. E.
1981 Seeds in Archaeological Sites: Sources and Some Interpretive Problems. *American Antiquity* 46:143–152.
1984 Native Food Plants of the American Southwest. *Masterkey* 58:3–8.
1985 *Social Adaptation to Food Stress: A Prehistoric Southwestern Example.* University of Chicago Press, Chicago.
1989 Prehistoric Diet in the Northern Southwest: Macroplant Remains from Four Corners Feces. *American Antiquity* 54:543–563.
1992 Earliest Plant Cultivation in the Desert Borderlands of North America. In *The Origins of Agriculture: An International Perspective*, edited by C. W. Cowan and P. J. Watson, pp. 121–141. Smithsonian Institution Press, Washington, D.C.
Morgan, L. H.
1965 [1881] *Houses and House-Life of the American Aborigines* (originally published as Volume IV of *Contributions to North American Ethnology*, [Government Printing Office, Washington, D. C.]). University of Chicago Press, Chicago.
Morris, D. H.
1990 Changes in Groundstone Following the Introduction of Maize into the American Southwest. *Journal of Anthropological Research* 46:177–194.
O'Shea, J. M.
1989 The Role of Wild Resources in Small-Scale Agricultural Systems: Tales from the Lakes and the Plains. In *Bad Year Economics: Cultural Responses to Risk and Uncertainty*, edited by P. Halstead and J. M. O'Shea, pp. 56–67. Cambridge University Press, Cambridge.
Patterson, W. A., III and K. E. Sassaman
1988 Indian Fires in the Prehistory of New England. In *Holocene Human Ecology in Northeastern North America*, edited by G. P. Nicholas, pp. 107–135. Plenum Press, New York.
Pearsall, D. M.
1988 Interpreting the Meaning of Macroremain Abundance: The Impact of Source and Context. In *Current Paleoethnobotany: Analytical Methods and Cultural Interpretations of Archaeological Plant Remains*, edited by

C. A. Hastorf and V. S. Popper, pp. 97–118. University of Chicago Press, Chicago.

Plog, F., G. J. Gumerman, R. C. Euler, J. S. Dean, R. H. Hevly, and T. N. V. Karlstrom
1988 Anasazi Adaptive Strategies: The Model, Predictions, and Results. In *The Anasazi in a Changing Environment*, edited by G. J. Gumerman, pp. 230–281. Cambridge University Press, Cambridge.

Powell, J. W.
1875 The Canyons of the Colorado. *Scribners' Monthly* 9: 523–537.

Powell, S.
1983 Stress and Success Among the Black Mesa Anasazi. In *Proceedings of the Anasazi Symposium 1981*, edited by J. E. Smith, pp. 107–111. Mesa Verde Museum Association, Mesa Verde, CO.
1988 Anasazi Demographic Patterns and Organizational Responses: Assumptions and Interpretive Difficulties. In *The Anasazi in a Changing Environment*, edited by G. J. Gumerman, pp. 168–191. Cambridge University Press, Cambridge.

Pyne, S. J.
1982 *Fire in America: A Cultural History of Wildland and Rural Fire*. University Press, Princeton.

Rautman, A. E.
1993 Resource Variability, Risk, and the Structure of Social Networks: An Example from the Prehistoric Southwest. *American Antiquity* 58:403–424.

Reinhard, K. J., P. R. Geib, M. M. Callahan, and R. H. Hevly
1992 Discovery of Colon Contents in a Skeletonized Burial: Soil Sampling for Dietary Remains. *Journal of Archaeological Science* 19:697–705.

Ronco, F., Jr.
1987 Stand Structure and Function of Pinyon-Juniper Woodlands. In *Proceedings—Pinyon-Juniper Conference*, compiled by R. L. Everett, pp. 14–22. General Technical Report INT-215. USDA Forest Service, Intermountain Research Station.

Rowlett, R. M.
1991a Ancient Grass Fires Detected by Thermoluminescence. *Archaeology in Montana* 32:29–32.
1991b Ceramic Thermoluminescence Response Effects after the La Mesa Forest Fire, Bandelier National Monument. *Archaeology in Montana* 32:49–56.

Schaber, G. G. and G. J. Gumerman
1969 Infrared Scanning Images: An Archaeological Application. *Science* 164:712–713.

Schlanger, S. H.
1991 On Manos, Metates, and the History of Site Occupations. *American Antiquity* 56:460–474.

Schlanger, S. H. and R. H. Wilshusen
1993 Local Abandonments and Regional Conditions in the North American Southwest. In *Abandonment of Settlements and Regions: Ethnoarchaeological and Archaeological Approaches*, edited by C. M. Cameron and S. A. Tomka, pp. 85–98. Cambridge University Press, Cambridge.

Smith, C. S.
1988 Seeds, Weeds, and Prehistoric Hunter-Gatherers: The Plant Macrofossil Evidence from Southwest Wyoming. *Plains Anthropologist* 33:141–158.

Stewart, G. R. and M. Donnelly
1943 Soil and Water Economy in the Pueblo Southwest. *Scientific Monthly* LVI:31–44;134–144.

Stewart, O. C.
1956 Fire as the First Great Force Employed by Man. In *Man's Role in Changing the Face of the Earth*, edited by W. L. Thomas, pp. 115–133. University of Chicago Press, Chicago

Sullivan, A. P., III
1986 Occupational Dynamics of a Small Kayenta Anasazi Settlement. In *Prehistory of the Upper Basin, Coconino County, Arizona*, edited by A. P. Sullivan, pp. 47–190. Arizona State Museum Archaeological Series No. 167. University of Arizona, Tucson.
1987 Seeds of Discontent: Implications of a 'Pompeii' Archaeobotanical Assemblage for Grand Canyon Anasazi Subsistence Models. *Journal of Ethnobiology* 7:137–153.
1992 Pinyon Nuts and Other Wild Resources in Western Anasazi Subsistence Economies. In *Long-Term Subsistence Change in Prehistoric North America*, edited by D. R. Croes, R. A. Hawkins and B. L. Isaac, pp. 195–240. JAI Press, Greenwich, CT.
1995 Prehistoric Alignments, Anthrosols, and Agriculture in the Grand Canyon Area. In *Applications of Pedo-Archaeology in the Old and New Worlds*, edited by A. C. Goodyear, J. E. Foss, and K. E. Sassaman. Anthropological Papers of the South Carolina Institute of Archaeology and Anthropology. University of South Carolina, Columbia.

Tainter, J. A.
1984 Foragers in the Southwestern Forests. In *Stage I Site Locational Modeling in the Southwestern Region*, edited by L. S. Cordell and D. F. Green, pp. 21–28. Cultural Resources Management Report No. 5. USDA Forest Service, Southwestern Region.

Thomas, W. L. (ed.)
1956 *Man's Role in Changing the Face of the Earth.* University of Chicago Press, Chicago.

Toll, M. S.
1988 Flotation Sampling: Problems and Some Solutions, with Examples from the American Southwest. In *Current Paleoethnobotany: Analytical Methods and Cultural Interpretations of Archaeological Plant Remains*, edited

by C. A. Hastorf and V. S. Popper, pp. 3–52. University of Chicago Press, Chicago.

Torrence, R.

1989 Retooling: Towards a Behavioral Theory of Stone Tools. In *Time, Energy, and Stone Tools*, edited by R. Torrence, pp. 57–66. Cambridge University Press, Cambridge.

West, N. E.

1984 Successional Patterns and Productivity Potentials of Pinyon-Juniper Ecosystems. In *National Research Council/National Academy of Sciences Developing Strategies for Rangeland Management*, pp. 1301–1333. Westview Press, Boulder.

West, N. E. and N. S. Van Pelt

1987 Successional Patterns in Pinyon-Juniper Woodlands. In *Proceedings—Pinyon-Juniper Conference*, compiled by R. L. Everett, pp. 43–52. General Technical Report INT-215. USDA Forest Service, Intermountain Research Station.

Wills, W. H.

1988 Early Agriculture and Sedentism in the American Southwest: Evidence and Interpretations. *Journal of World Prehistory* 2:445–488.

1992 Plant Cultivation and the Evolution of Risk-Prone Economies in the Prehistoric American Southwest. In *Transitions to Agriculture in Prehistory*, edited by A. B. Gebauer and T. D. Price, pp. 153–175. Westview Press, Boulder.

Wills, W. H., P. L. Crown, J. S. Dean, and C. G. Langton

1994 Complex Adaptive Systems and Southwestern Prehistory. In *Understanding Complexity in the Prehistoric Southwest*, edited by G. J. Gumerman and M. Gell-Mann, pp. 297–339. Addison-Wesley Publishing Company, Reading, MA.

Winter, J. C. and P. F. Hogan

1986 Plant Husbandry in the Great Basin and Adjacent Northern Colorado Plateau. In *Anthropology of the Desert West: Essays in Honor of Jesse D. Jennings*, edited by C. J. Condie & D. D. Fowler, pp. 117–144. University of Utah Anthropological Papers No. 110. University of Utah Press, Salt Lake City.

Winterhalder, B.

1990 Open Field, Common Pot: Harvest Variability and Risk Avoidance in Agricultural and Foraging Societies. In *Risk and Uncertainty in Tribal and Peasant Economies*, edited by E. Cashdan, pp. 67–88. Westview Press, Boulder.

Wright, H. A. and A. W. Bailey

1982 *Fire Ecology*. John Wiley and Sons, New York.

Wright, H. A., L. F. Neuenschwander, and C. M. Britton
1979 *The Role and Use of Fire in Sagebrush-Grass and Pinyon-Juniper Plant Communities: A State-of-the-Art Review.* General Technical Report INT-58. USDA Forest Service, Intermountain Forest and Range Experiment Station.

Timothy A. Kohler*[*] and Carla R. Van West[**]
*Department of Anthropology, Washington State University, Pullman and Santa Fe Institute, 1399 Hyde Park Road, Santa Fe, NM 87501
**Crow Canyon Archaeological Center, Cortez, and Statistical Research, Inc., Tucson, AZ

The Calculus of Self-Interest in the Development of Cooperation: Sociopolitical Development and Risk Among the Northern Anasazi

INTRODUCTION

One clear trend that can be discerned in the last 15 years of otherwise-rather-protean Southwestern archaeology is a growing recognition that at any given time, demographic, productive, and organizational strategies were quite variable, even within comparatively small regions. Within a general trend of increasing economic intensification, strategies and tactics of production (Leonard and Reed 1993) could change quickly and in ways that are not, apparently, easily predictable. Perhaps it is part of the process of normal science to begin by emphasizing the most obvious patterns in time and space—as exemplified for the Anasazi by the Pecos classification—and only then, to focus on understanding the reasons for the variability that remains, unexplained and initially even undescribed, around those modes. In any case, we have moved from an almost complete focus on mean, or normative, behavior to a greater concern with variability around those norms.

Part of this process of elaborating normative thought has been to reexamine the archaeological record for information on how climatic or environmental *variability* (as opposed to average conditions) has affected prehistoric behavior. Of the great deal of work done in this area, that of Braun and Plog (1982) is perhaps the most widely cited. Their central thesis is that "*sustained increases in the intensity of integration will occur as concomitants of sustained increases in region-wide uncertainty or risks arising from environmental change*"[1] (emphasis in original). Their focus is on *regional* connectedness (presumably including interaction and cooperation), measured especially through stylistic traditions; environmental risks are measured primarily through the proxy of increasing local populations, which are held to increase risk through restricting mobility after a certain level of population is reached. In a similar vein, Upham (1984) has argued that *food exchange* among large late prehistoric sites in *local* clusters is related to the management of production risk in high-risk/high-diversity environments. Implicit in such a model is that food exchange should increase in times of increasing risk (we will be using risk here to mean variability in agricultural production, as measured by the standard deviation, rather than as the probability of falling below some acceptable level of income). In general these approaches assume some sort of positive linear relationship between increasing variability and increasing "integration," usually making a further connection between increasing connectedness, or integration, and increasing opportunities for emergence of hierarchical sociopolitical systems.

The present chapter builds on this earlier work to formulate a new framework for analyzing and predicting the behavior of producers in various contexts of productive shortage and abundance. We begin by assuming that nearly all agricultural production among the northern Anasazi was at the household level. By drawing on microeconomic theory, evolutionary ecology, and recent studies in the behavioral ecology of foragers, we then develop a context to predict when networks of food exchange among households ought to appear, based solely on the self-interest of the households involved. Data on production variability in the Mesa Verde Region between A.D. 901 and 1300—recently developed by Van West (1994)—are examined in this context, and these sections draw heavily on Van West and Kohler (1995). We then try to argue that networks of food sharing provide a foundation for more general elaboration of sociopolitical complexity in cases where production provides a surplus. In the final sections of this chapter we examine the implications of this work and point to some of the fruitful avenues for research that it suggests.

[1]Their argument is meant to apply only to situations which are not severe enough to endanger the existence of the network itself (Braun and Plog 1982:508). We will try to make the case that systems of interaction, coordination, and sharing exist on several levels of spatial and social generality. Not all of these are subject to collapse, but those at the furthest reaches of social and spatial distance are constantly at risk and do, in fact, collapse frequently.

UTILITY FUNCTIONS AND RISK SENSITIVITY

As archaeologists have been searching for such linear relationships between risk and behavior over the last 12 years, studies of animal foraging systems during the same period have moved in different directions. An exemplary early study (Caraco et al. 1980) drew on utility theory to explain why juncos sometimes prefer a risky foraging strategy (involving a variable number of seeds), and sometimes a risk-free foraging strategy (involving a fixed number of seeds), when the long-term payoffs from both were the same. Utility theory was developed in economics in the mid- to late nineteenth century, although Daniel Bernoulli, in the 1730s, was already postulating that the value of an additional dollar to a person was inversely proportional to his current wealth (Rima 1972:182).

> The concept of marginal utility expresses the subjective value or want-satisfying power of an additional unit of a given good to a particular user. The importance an individual attaches to an additional unit of a particular good depends in part on its relative scarcity. The larger the supply of a given commodity, the smaller will be its relative significance at the margin (Rima 1972:182–183).

A *utility function* relates some amount of goods (on its abscissa, or x-axis) to some psychological utility or value for those goods to a consumer (on the ordinate, or y-axis). Caraco et al. (1980) noticed that when birds were deprived of food, they were more likely to choose a variable reward, whereas they favored a fixed reward when their energy budget was positive. Deprivation (or satiety), they concluded, changed the shape of the utility function that guided junco foraging "decisions" so that it was not linear in form (see also Krebs and Kacelnik 1991; Stephens 1981). Decisions that systematically discriminate between different strategies that have the same expected (mean) value but differing degrees of variance are said to be risk-sensitive (Kaplan et al. 1990; Stephens 1990).

Consider the utility function graphed in Figure 1(a). It represents one possible relationship between total output of a foraging or production system as measured in units such as bushels or kilocalories (on the x-axis) and the *value* or *utility* of that production to the producer (on the y-axis). (An equivalent analysis, in the language of behavioral ecology, could be made by considering fitness outcomes rather than utility on the y-axis.) If the relationship were a straight line, then the difference in value between, say, 10 and 20 bushels of corn would be the same as that between 30 and 40. There is, however, reason to expect that this relationship is often of the general sigmoid form graphed here (see Smith 1988:236; Smith and Boyd 1990:170–172). By this we do not suggest that we can define a specific curve for this (or any) archaeological context, but our conclusions will strengthen the conjecture that the appropriate curve belongs to the general, sigmoid-shaped family of functions. In such a relationship, the value of marginal production increases more

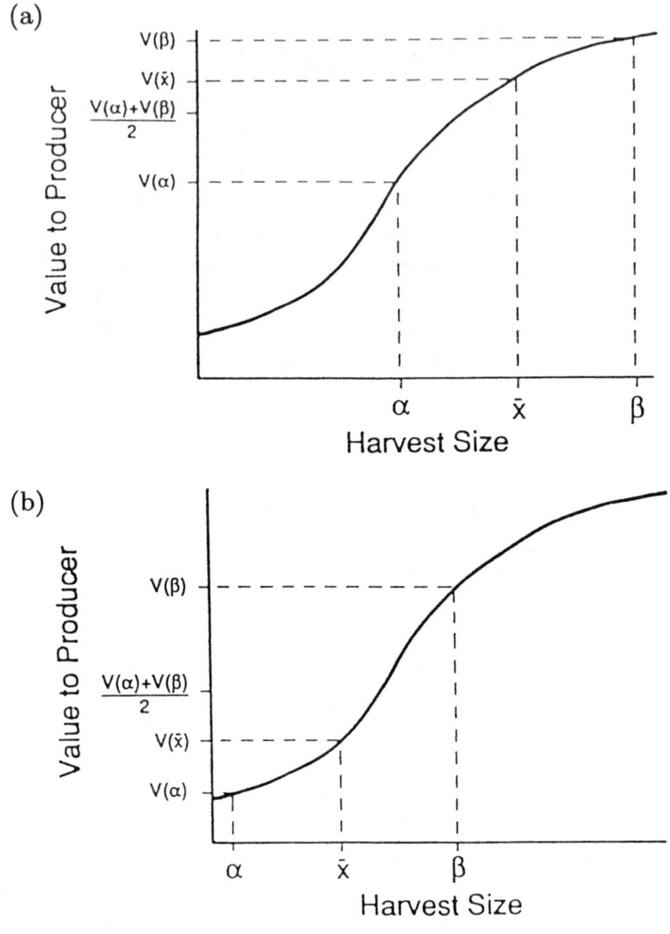

FIGURE 1 (a) Good-year economics. Units of production on the x-axis; units of value to the producer of a certain level of harvest on the y-axis. The variability notated by α and β could represent different plots in the same year, or different years in a series of generally good years. $V(\bar{x})$ represents the average value realized after pooling; $[V(\alpha) + V(\beta)]/2$ represents the average value realized by not pooling. After Smith (1988). (b) Bad year economics. In contrast to good years, in bad years $V(\bar{x})$ (the expected value of pooling) is less than $[V(\alpha) + V(\beta)]/2$ (the expected value of not pooling).

quickly than production itself when few units are being produced, but more slowly when many units are being produced. So the difference in value between 10 and 20 bushels of corn may be either more—or less—than the difference between 30 and 40 bushels, depending on how the inflections on the utility function interact with these production levels on the x-axis. Slightly more formally, the sigmoid curve graphing total utility starts off with zero slope, has an increasing first derivative (which is the marginal utility) until the second derivative reaches zero, and then a diminishing marginal utility as the first derivative asymptotically approaches zero. One important emphasis of this chapter is that the household decision to share, or to hoard, should be affected by whether household production is above, or below, the major inflection point where the second derivative equals zero, beyond which marginal utility begins to decrease.

In the lower left-hand portion, or upward-concave portion, of the total utility curve, then, the marginal utility of another bushel is always greater than the last. The flat, initial segment of the curve may be thought of as indicating the presence of an "outside option"; under extremely adverse conditions—below some threshold required to meet reciprocal obligations or avoid starvation, or below the minimum threshold required to make a living as a farmer—defection into the countryside to pursue a foraging strategy, or to join a different group, must have been a possibility at most times. The diminishing marginal value segment of the curve (concave downwards, in the upper right-hand portion of the graph) is based on a great deal of economic research in the tradition discussed by Rima above. In the present case, for subsistence farmers, we presume that an additional bushel of maize in a very large harvest is not as valuable as an additional bushel of maize in a moderate harvest, since very large harvests present storage problems; since maize degrades to some extent during storage; and since transportation for exchange is not free. Yet very large harvests probably occur, given lucky combinations of planting and harvesting conditions, even if farmers plant only enough to get adequate harvests under average conditions. Sebastian (1991) makes the interesting argument that the ninth-century pithouse-to-pueblo transition marks the development of a strategy of *overproduction* (in average and good years) and an emphasis on storage, rather than foraging, to get through lean years.

An important characteristic of these curves is the way in which values of means taken on the x-axis differ from the mean values of the unaveraged yields taken on the y-axis. If mean production falls within the diminishing marginal value segment of the curve (as in Figure 1(a)), then the mean value of having chosen a risky gain (simplified here as $[V(\alpha) + V(\beta)]/2$ is *lower* than the mean value of having chosen a risk-free gain ($V[\bar{x}]$).[2] At high production levels, each household will optimize

[2] Imagine α as the harvest of some household in one year, and β as the harvest of the same household in the following year. (α could also be the production of a household in year 1, and β the production of another household in the same year.) If—in the first case—the household shares its production with no other household, then after two years, its expected value realized from both harvests will be $[V(\alpha) + V(\beta)]/2$. (Here the value for each year's production is realized on the y-axis and then the average value over two years is calculated.) On the other hand, if the harvest was generally shared by households within a village *before* its value was "realized" (consumed), then the expected value for each household would be the average ($V[\bar{x}]$) across all the households in the village. In this case, the mean is calculated on the production axis, and then its value is "realized" on the y-axis.

the value of its production by consuming at the mean production level (say, for its village) rather than averaging the value of consuming randomly-varying harvests through hoarding and storage (because, in good years, $V(\bar{x}) > [V(\alpha)+V(\beta)]/2$). In other words, households should avoid risk during a series of good years. Conversely, if we slide the $\bar{x}$ and variance indications α and β well to the left on the x-axis, the mean value of production is increased by consuming the average of the stochastic variability (because, as in Figure 1(b), $V(\bar{x}) < [V(\alpha) + V(\beta)]/2$). In bad years, households should be risk-seeking in their production/consumption strategies.

There are two major dimensions to variability in agricultural production for the Anasazi. First, holding space constant, there is variability from year to year in the production of any given plot due to climatic fluctuation (and other factors); second, within any year, there is spatial variability in the productivity of various agricultural plots. Many behaviors have been linked to attempts to avoid these risks. Two important strategies are storage, to buffer the year-to-year variability, and pooling food among producers, to buffer spatial variability within any year.[3] Since stored food may also be subject to pooling, and pooled resources may not be immediately consumed, it might be expected that these conceptually distinct responses to temporal and spatial variability might in practice be correlated. The terms "pooling" and "sharing" are used as synonyms here to encompass both the restricted sharing and the more general pooling of resources distinguished by Hegmon (1989). Although the precise sharing rules in place are of great interest, all we need to assume about those rules at this point is that (1) the system provides for those producing above the local mean to give more to those who produced below the mean than vice versa; and (2) that reciprocal obligations are recognized and can generally be met. The number of sharing households necessary to achieve dramatic reductions in income variance is related to the degree of correlation in production among households, and may in practice be fairly small (Hegmon 1989; Winterhalder 1986).

TO SHARE OR NOT TO SHARE

The model to be examined here is quite simple, although some of its predictions are not intuitively obvious. The decision variable to be analyzed is whether or not a household should pool its agricultural production prior to consumption. A number of assumptions are made.[4] In periods characterized by relatively high mean

[3]Keeping many plots is another obvious option (Hegmon 1989; this volume). Whereas we do not deny that this may have been of some importance, as it was clearly practiced in historic times, it would be difficult for a household to monitor many plots during the key harvest period. This option is probably more attractive in landscapes with relatively little wild game, as in the Medieval English situation discussed by Winterhalder (1990).

[4]These include: (1) the shape of the utility function; (2) that each household seeks to optimize the utility from its consumption; and (3) that local household harvest rates are not perfectly correlated across years.

production, behaviors involving pooling of production ought to be attractive, both for the prodigious households (since the value of production they give away through sharing in a good year will be less to them than the value of the production they expect to receive through sharing in a bad year) and for the less fortunate (for the same reasons).[5] If these same periods are also subject to relatively high year-to-year fluctuation (we will be measuring temporal and spatial variability below in terms of their standard deviations), the difference between the value of the mean post-pooling consumption rate ($V[\bar{x}]$) and the value of the mean nonpooling consumption rate ($([V(\alpha) + V(\beta])]/2$) is accentuated (compare Figures 2(a) and (b)). So periods with high mean production, coupled with high annual fluctuation in production, ought to be especially favorable for the development of pooling behaviors. Finally, and for the same reasons, periods having high spatial variability in production should also tend to favor risk-averse (pooling) behaviors. As long as these conditions persist, and as long as the *distribution* of yields (*not* the yields themselves) is more or less similar for all households, the system of sharing will seem to be to the advantage of all, and ought to persist.[6]

On the other hand, periods when mean production is quite low ought to discourage sharing (and favor defection from any ongoing system of sharing). (The term "defection" is borrowed from game theory [e.g., Axelrod 1984:8] as shorthand for the decision not to go along with some system of cooperation.) This is because

[5]Smith and Boyd (1990) show how an analogous system among hunter/gatherers can profitably be modeled as a two-person or n-person game, having payoffs to participants structured according to the Prisoner's Dilemma (PD). (Sigmund [1993:180–206] provides an excellent introduction to the study of this social dilemma, in which the rational strategy is to defect, but with undesirable outcome that if both players defect, they receive payoffs that are less than if both cooperate.) For the one-shot PD game, the dominant strategy is to defect (not share). Two obvious ways out of the dilemma, both eminently applicable to the Puebloan situation, are strong social sanctions against defection, and/or an "iterated structure" to the game which makes it very likely that current players will meet again (Axelrod 1984:9–21; see also Gumerman and Dean 1989). In a stable village setting, the future must loom large in this manner. However, we will argue that there may not have been strong social sanctions in place against defection until about A.D. 1300 throughout much of the northern Southwest. For the system analyzed here, the average payoff for cooperation will be higher than the average payoff for defection only in times of relatively high production (as in the PD); thus, in periods of low production, this is not a PD game.

[6]In this chapter we stress climatic-economic reasons why such systems break down, but it is also possible to see, in this formulation, how the system would be endangered if some households *continually* produced more than others. With no expectation for receipt of valuable reciprocal goods in bad years (since there are no bad years), such households could defect (in the absence of very strong social sanctions). This situation is in fact covered by the model, to the extent that we predict sharing to be less valuable in times of little temporal variability in yields. However, the example illustrates that the distribution of yields over time may not be the same for all households in a village, especially given a developed system of land tenure with field ownership at the household, lineage, or clan levels. It also illustrates that defection from a reciprocal system would not necessarily originate in the less-productive households.

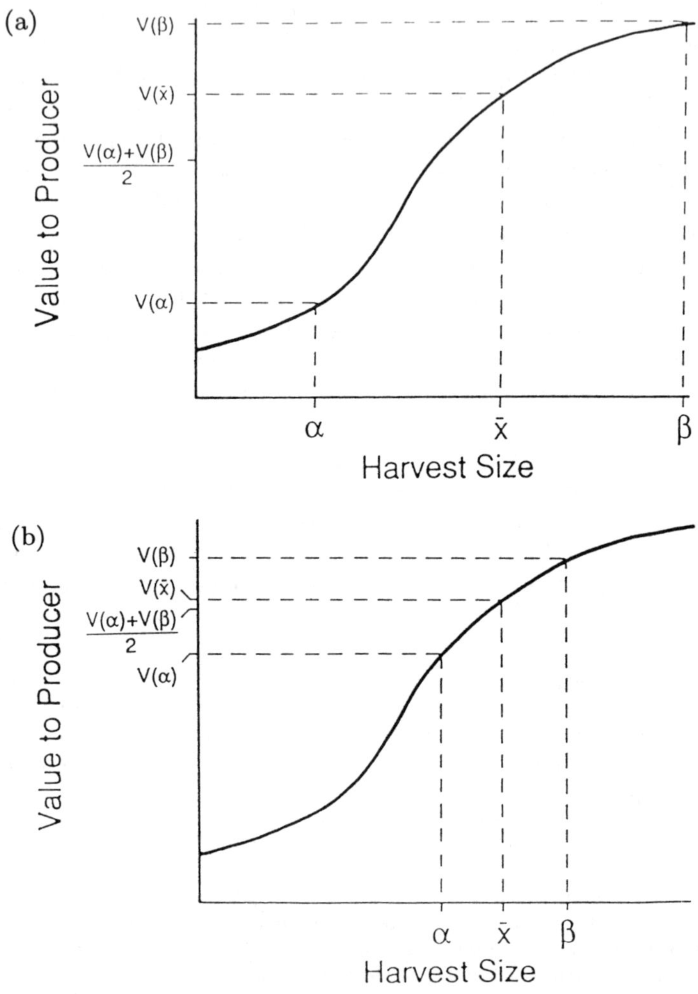

FIGURE 2 In good years, increasing production variance (through time or space) increases the relative value of pooling when average production is held constant. (a) A series of years with a relatively high mean production and relatively high variance results in relatively large differences between the expected value of sharing versus not sharing. (b) A series of years with the same mean but relatively low variance results in less difference between the expected value of sharing versus not sharing. Notation as in Figure 1.

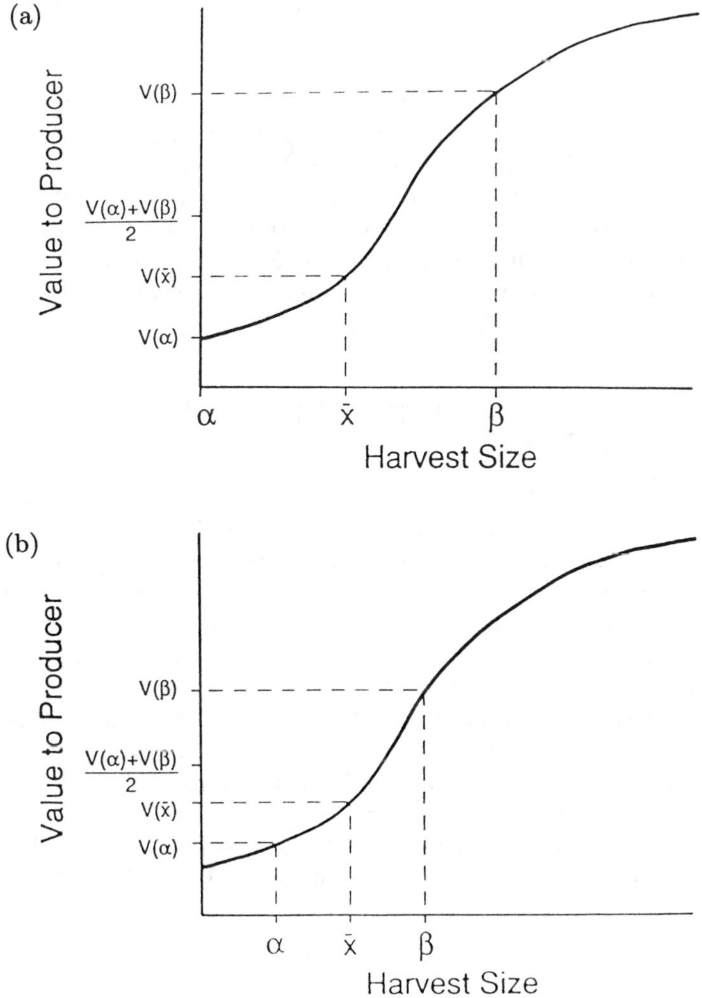

FIGURE 3 In bad years, increasing production variance (through time or space) increases the relative value of not pooling when average production is held constant. (a) A series of years with a relatively low mean production and relatively high variance results in relatively large differences between the expected value of sharing versus not sharing. (b) A series of years with the same mean but relatively low variance results in less difference between the expected value of sharing versus not sharing. Notation as in Figure 1.

the value (on the y-axis) of the mean risky (nonpooling) consumption is higher than the value of the mean risk-free (i.e., post-pooling) consumption. Moreover—and this point seems to go against traditional archaeological intuition—high temporal or spatial variability in periods of low mean production will exaggerate the difference between the values of the mean risky and the mean risk-free consumption rates. In periods of low mean production, the relative value of defection compared with sharing is greatest when temporal or spatial variability is *greatest* (compare Figures 3(a) and (b); see Hegmon 1989:93 for a related point, expressed in the currency of risk reduction rather than utility maximization). Another way to say all this is that when yields are generally poor, and have been for some time, the prodigious (but still hungry) household is discouraged from sharing by the fact that the production it gives away (in a slightly less lean year) is greater in value than the production it can hope to receive (in a year that is even worse).

SUMMARY OF THE MODEL FOR PREDICTING DEVELOPMENT OF FOOD SHARING AMONG HOUSEHOLDS

Pooling of food (a form of risk-averse behavior, and perhaps a critical element of cooperative behavior in general) is most likely to develop in periods of high mean productivity, high variability in productivity from year to year, and great differences in productivity across space. In periods characterized by low mean productivity but high temporal and/or spatial variability, pooling is not in the best interests of the participants, and is expected to break down, if present, or not to develop. Because we cannot directly observe utility functions for any society, especially a prehistoric one, it is comforting to learn that Sebastian (1991:111) has marshalled ethnographic evidence to show that small-scale agricultural societies do tend to share in good times, and hoard in lean times, a behavior which makes sense only if their implicit utility functions resemble those assumed here. In the next two sections we discuss how periods that meet these conditions are defined, and how "pooling behavior" (and its demise) might be recognized in the archaeological record.

METHODS FOR EVALUATING MODEL PERFORMANCE

IDENTIFYING PERIODS OF INTEREST

Because the shape of the utility function entails that the expected value of pooling will exceed the expected value of not pooling only when the mean production is relatively high, it was first necessary to identify periods of relatively high mean production in the record. Such periods should be relatively long in order to have some chance of being able to recognize them in the archaeological record, and to allow decisionmakers to have some idea as to the relative payoff for pooling versus defection in order to be able to make decisions on that basis. This would probably be impossible on the basis of very few years with given conditions.

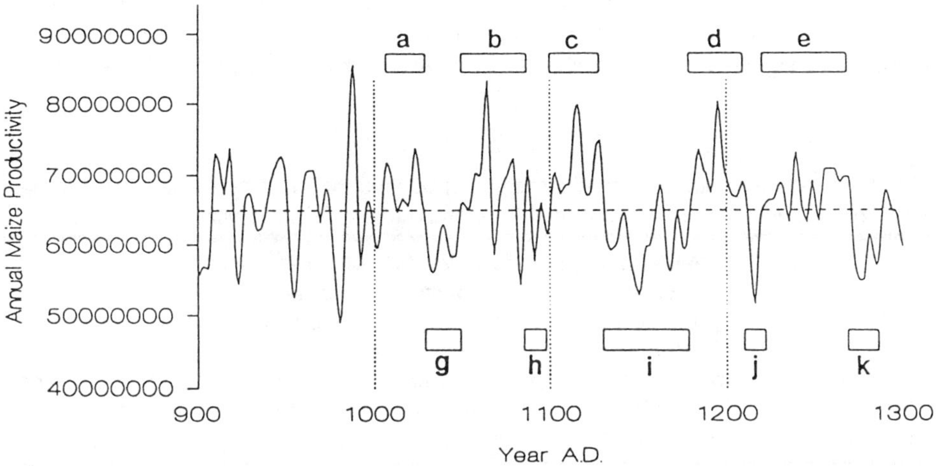

FIGURE 4 Smoothed annual estimates of total maize productivity in kilograms. Periods identifiable as favorable for, or unfavorable for, pooling are identified by bars above and below the series, respectively.

Periods with relatively high mean production were found by smoothing Van West's (1994:133) estimates for total maize production for a study region encompassing about 1500 km.^{2} in Southwest Colorado. These estimates were developed for each year from A.D. 901–1300, at a spatial scale of 4 ha., and are sensitive to soil depth classes, elevation, and soil moisture (as retrodicted through estimates of Palmer Drought Severity Indices, calibrated from the relationship between modern temperature and precipitation data and recent tree-ring indices). Van West (1994) describes in detail how these paleoproductivity estimates are derived, and the environments and prehistory of the study area. The smoothed series of production values is displayed in Figure 4. This series appears to encompass five periods (from 24 to 50 years in length) of relatively high average production, identified by solid bars above the graph, and five (somewhat shorter) periods of relatively stable low means, identified by bars below the graph. The periods of below-normal production, expected to be unfavorable for pooling, range in length from 10 to 50 years; many of the unfavorable periods are rather short. This left 117 years (A.D. 901–1005 and 1289–1300) which were not part of either a stable high or low period and which are considered neutral in terms of the model.

The next step was to compute measures of temporal and spatial variation in production within each of these defined periods. This is necessary because the model predicts that pooling will be most attractive in periods with relatively high means that also exhibit high temporal and spatial variability in production. Temporal variability was measured here as the standard deviation around the mean production (per hectare per year) for each period. Spatial variability within each year was likewise measured by the per hectare standard deviations, and these annual

TABLE 1 Periods with differential advantages for pooling, ordered by median rank for mean total maize productivity and measures of temporal and spatial variability.

Period[1] in Years A.D. (median ranks)	Mean annual maize productivity during period x 100 (ranks within group)	Standard dev. for annual maize productivity during period x 100 (rank within group)	Mean standard deviation in production across region during period (rank within group)	Value of pooling as predicted by model
[c] 1100–1129 (1)[2]	71,356 (1)	13,702 (3)	305 (1)	strongly positive
[d] 1180–1211 (2)	70,797 (2)	14,893 (2)	303 (2)	|
[a] 1006–1029 (3)	68,118 (3)	13,683 (4)	299 (3)	|
[b] 1049–1088 (4)	68,109 (4)	16,000 (1)	294 (5)	|
[e] 1222–1271 (5)	66,418 (5)	11,772 (5)	296 (4)	weakly positive
all 117 years not included in a favorable or unfavorable period	64,666	14,284	289	approx. neutral
[h] 1089–1099 (1)	59,607 (1)	10,961 (3)	275 (1)	weakly negative
[g] 1030–1048 (3)	59,046 (3)	10,687 (1)	277 (3.5)	|
[i] 1130–1179 (3.5)	59,433 (2)	12,130 (5)	277 (3.5)	|
[j] 1212–1221 (4)	58,333 (4)	11,866 (4)	276 (2)	|
[k] 1272–1288 (5)	58,033 (5)	10,797 (2)	280 (5)	strongly negative
901–1300[3]	64,925	13,937	290	—

[1] The bracketed letters in front of each period identify the periods in the plot of smoothed regional production values (Figure 4).

[2] The median ranks of periods in this analysis differ slightly from those in Van West and Kohler (1995), where measures of temporal and spatial variability were based on residuals from regressions against productivity measures.

[3] Means for the entire series, for comparison.

measures were averaged for all the years within each defined period to form a measure of average spatial variability in each period.

For each period, the mean production and these measures of temporal and spatial variation are displayed in Table 1. (The accuracy of these figures can be roughly estimated for the historic period, but not for the prehistoric period, and it is likely that we present more significant digits in this table than is warranted.) In the first column of this table, the periods are ranked according to the best estimate of the overall attractiveness of pooling. For the set of periods in which pooling is expected to develop, this ranking is achieved by first ranking the scores in columns 2, 3, and 4 from high to low (with a rank of 1 assigned to high positive scores). Then each period is assigned an overall rank by taking the median of these three ranks. Thus, the ranks for period [c] are 1, 3, and 1, yielding a median of 1,[7] the highest rank for any period. We used ranks, rather than standardizing these three values as z-scores and taking their mean, because of some disjunction between what we would like to measure (the achieved production per household, given some particular distribution of population and fields in any given year) and what we are actually measuring (an estimate of potential production across the entire landscape).[8]

The periods in which the expected value of defection is greater than that of sharing were then ranked according to the same logic. The greatest rewards for not sharing are connected with low mean production (the lowest is assigned a rank of 5 in Table 1, column 2), high temporal variation (the highest is assigned a rank of 5 in column 3 of the same table), and high spatial variation (the highest is assigned a rank of 5 in column 4). Then the median of these three ranks was used to assign to each period an overall attractiveness for sharing behavior. For example, period [k] with ranks of 5, 2, and 5 received a median rank of 5, identifying it as the least favorable period for pooling in the 400-year record. This is the period from A.D. 1272 through 1288.

IDENTIFYING POOLING BEHAVIOR IN THE ARCHAEOLOGICAL RECORD

To examine the predictions of this model we must be able to monitor the growth and demise of cooperative behavior in the northern Anasazi Southwest, particularly those cooperative behaviors that might involve food sharing. The vast literature on integration, reciprocity, redistribution, political complexity, and aggregation is all germane in attempting to identify what these behaviors might be. Our tactic in choosing measures that ought to be involved with food sharing was to choose as many indices as possible in the full realization that none may be a pure measure of

[7] For the case of three periods with one tie, the median will equal the mode.

[8] This disjunction is probably greatest when population levels are relatively low, and farmers have more freedom to select parcels of land that they judge likely to do well in the following year, based on previous experience. When population levels are higher, the variability in achieved production probably more closely mirrors that of the larger landscape.

TABLE 2 Initial test of the Pooling Model.

	Predicted High Positive ⟵ Strength of Pooling Advantage ⟶ High Negative										
	1100-1129	1180-1211	1006-1029	1049-1088	1222-1271	Neutral years	1089-1099	1030-1048	1130-1179	1212 1221	1272-1288
Expectations for Periods in which Pooling is Expected:											
Growth/aggregation at site level (e.g., more rooms indicate more residents)	++	+	+	+	+++	?	?	?	+	?	-
Growth/aggregation at community level (e.g., more sites indicate more members)	++	+	+	+	++	?	?	?	?	?	-
Great Kivas	7	2	?	1	5	?	?	?	?	?	0?
Reservoirs	+	+	?	?	++	?	?	?	+	?	-?
Great Houses	10	?	?	1	3	?	?	?	+	?	0
Roads	5	1	0	?	1	?	?	?	+	?	0
Enclosing Walls/Interior Plazas	1	?	0	0	6	?	?	?	?	+	0?
Triwalled and Biwalled Structures	1?	0	0	0	5	?	?	?	?	?	0?
Foundation of hamlets associated with the establishment/growth of aggregates	++	+	++	+	+++	?	?	?	-	?	-
Expectations for Periods in which Defection is Expected:											
Breakup of aggregates	?	?	?	?	?	+[1]	?	?	+	?	+
Overall strength of evidence for pooling (rank relative to other periods)	1	3	5	4	2	not ranked	9	9	6	7	9

[1] This refers to the breakup of the mid-ninth-century Pueblo I villages (as in the Dolores Archaeological Project area) which may take place slightly before A.D.900.

the dimension. We consider all to be at least weakly involved with the dimension of interest.

The measures selected are listed in the first column of Table 2. More discussion of how these facets of the archaeological record might be involved with food sharing can be found for aggregation at the site level in, for example, Glassow (1977:206); for growth and aggregation at the community level in Orcutt et al. (1990) and Sebastian (1991); for great kivas and triwall structures in Plog (1974:127); for reservoirs in Haase (1985); and so forth. Ford (1972) provides a general perspective on the importance of the movement of food in contemporary Tewa ritual and society. Our underlying premise is that all of the cooperative behaviors implied by these collective features—but especially aggregation itself—are built on the foundation of food sharing and provide a theater where continual face-to-face contacts reinforce mutal obligations and make free-riding easier to detect. Other measures of increased interaction, such as higher intraregional rates of exchange of regionally produced ceramic and lithic materials, should also be expected in periods in which regional production potentials favored development of pooling. Unfortunately, we know of no studies that describe the volumes of any intraregional flows of materials with enough temporal precision to be useful in testing the present model.

Table 2 lists only one "positive" test implication for the periods in which sharing is expected to break down—the dissolution of aggregated sites. Of course, we also expect to see no evidence in these periods of the behaviors we connect with sharing.

The temporal precision of our expectations exceeds the temporal precision of a good deal of the archaeological record. For this reason the tree-ring dated sites in the study area (Van West and Kohler 1995:Table 4) are especially valuable for testing the expectations. We add to the group of tree-ring dated sites another group of sites for which probable peaks of occupation can reasonably be derived from ceramic materials, and tabulate those items of public architecture from this larger set of sites that we wish to use as indices of increased sharing of resources (see Van West and Kohler 1995:Table 5). Finally, in Table 2, these data are tabulated against the periods identified under our model (in Table 1) as either rewarding cooperative food sharing, or rewarding defection.

RESULTS

The general pattern of the record is strongly in the directions anticipated by the model. The period between 1100 and 1129—in which the expected value of cooperative behaviors is highest—coincides with the local buildup of the "Chacoan system." The mid-1200s, when the "terminal" aggregation takes place at canyonhead sites such as Sand Canyon, is also, correctly, predicted to be a time when cooperative behaviors are expanding or stable. The breakup of the local Chacoan-related system (between 1130 and 1179) and the final abandonment of the region (in the 1270s or

1280s) are likewise found in those periods in which defection ought to be advantageous, according to the model. Finally, although we did not include it as a formal expectation for periods of defection in Table 2, cannibalism perhaps represents the ultimate in negative reciprocity. The occupations of the four sites within or near our study area for which White (1992) finds strong evidence for cannibalism generally coincide with periods for which defection is predicted. The Grinnell site (A.D. 1135–1150, White 1992:376) is occupied entirely within period [i]; 5MT3 at Yellow Jacket (A.D. 1025–1050, White 1992:378) is occupied almost entirely within period [g]; 5MTUMR-2346 is poorly dated but the probable occupation in the first half of the twelfth century generously overlaps period [i]. Perhaps least likely to be in conformity with our prediction is the cannibalism at 5MT1 in the Yellow Jacket complex, probably dating to the late A.D. 900s and early 1000s, apparently overlapping neutral, favorable, and unfavorable periods, although possibly terminating in unfavorable period [g].

Some of the apparent failures in prediction may reflect weaknesses in our ability to date the archaeological record precisely. We suggest that many of the "+" signs in the 1130–1179 period in fact pertain to sites actually belonging to the immediately preceding period of more favorable conditions. Three of the periods in which we predict defection (1030–1048, 1089–1099, and 1212–1221) are simply too short to identify with confidence in the record.

How does this model perform relative to some obvious rivals for predicting the timing of increase in cooperative behaviors? In Table 3 we examine two such models—one in which simple abundance (high productivity) is used as a predictor of cooperative behaviors (column 4), and a second, "risk-buffering" model, similar to that espoused by Braun and Plog, where periods of greatest temporal and spatial variability are predicted to have most evidence for cooperative behaviors (column 5).

The performance of all three models is quite good, and probably undifferentiable, given the vagaries of the data. The correlation between the predictions of our model (labeled the "risk-sensitivity model" in column 3 of Table 3) and the archaeological record as we read it, is 0.60 as measured by Kendall's Tau-b (τ_b); between the simple abundance model, and the record, 0.60; and between the "integration as risk-buffering" model, and the record, 0.71.

In fact the predictions of all three models are very similar, as can be appreciated from the last row of Table 3, which displays the correlation coefficients among the predictions. In this landscape, the spatial variance in production across the study area in any year is highly correlated with the mean annual (per hectare) productivity. Areas of thin rocky soils in the study area do poorly almost all the time, whereas production in areas of deep loess soils is very responsive to recent and current precipitation. It should also be noted that *periods* in which mean productivity is relatively high tend to display relatively high year-to-year variability, probably because of the mean/variance effect, in which high means entrain high variances, but also because there is a floor below which productivity rarely falls, so that

TABLE 3 Performance of competing models for development of cooperative behaviors in the archaeological record.

Period (A.D.)	Deg. of integration seen in record (rank based on data on Table 2[1])	Deg. of integration predicted by risk-sensitivity model (Table 1)	Deg. of integration predicted by risk-abundance	Deg. of integration predicted by buffering model[2]	Deg. of integration predicted by risk-buffering model with abundance removed[3]
1100–1129	1	1	1	1.5[4]	8
1222–1271	2	5	5	5	6.5
1180–1211	3	2	2	1.5[4]	4
1049–1088	4	4	4	3	5
1006–1029	5	3	3	4	2.5
1212–1221	6	9	9	7.5[4]	2.5
1130–1179	8.5[4]	8	7	6	6.5
1030–1048	8.5[4]	7	8	9	9
1089–1099	8.5[4]	6	6	10	10
1272–1288	8.5[4]	10	10	7.5[4]	1
τ_b correlation coefficients with column 2 data		0.60	0.60	0.71	-0.07
τ_b correlation coefficients with "risk-sensitivity" model			0.96	0.61	-0.16

[1] In making this ranking, we assumed, as suggested above, that the '+' marks in the 1130–1179 column of Table 2 are due primarily to problems in dating sites that are in fact slightly earlier; that is why the ranks here differ slightly from those in Table 2. Only more research at the questionable sites could settle this problem.

[2] This ranking was calculated after removing the linear relationship of the measures of temporal and spatial variability (columns 3 and 4 of Table 1) with a measure of annual maize productivity (column 2, Table 1) through regression. The resultant residuals are termed measures of "relative" temporal and spatial variation by Van West and Kohler (1995) and are displayed in columns 4 and 5 of their Table 3.

[3] Calculated by ranking the periods based on the average of the ranks for temporal and spatial variability (using data from columns 3 and 4 of Table 1), taking ties into account as appropriate.

[4] Ties.

periods of relatively low productivity (with few years departing radically from that floor) automatically display low year-to-year variance. Taken together, these two effects dictate that periods of high temporal and spatial variability (which according to the risk-buffering model ought to favor increased integration) are almost automatically periods of high abundance, *and* periods when the calculus of self-interest discussed above would predict food sharing. That the risk-buffering model performs well because of this effect, and not because of variability per se, can be seen in the last column of Table 3, where we subtract the linear dependence that our measures of temporal and spatial variability have on raw production. So recast, this model performs very poorly against the archaeological data. In future comparisons of these models, it would be desirable to select more carefully the spatial frame so that areas of relatively low productivity are not included; this would minimize the mechanical correlation between high productivity and temporal and spatial variability in productivity. However, it is inescapably a characteristic of these landscapes that these factors are correlated to some extent.

In assessing the performance of the risk-sensitivity model, it is important to remember that the household-level decisions that it predicts are being made on the basis of perceived *per household* values. Thus, the regional productivities (by which we predict the household-level decisions) ought to be corrected for regional population sizes. We have not been able to do so, although such corrections would be possible for some subsets of the study area. With such a (negative) correction to income (on the x-axis of our utility functions) the relative value of pooling (on the y-axis) would likewise tend to decrease, and this effect should be more marked in times of high population than in times of low population. Therefore, when populations are high, a given level of production and degree of spatial and temporal variability should result in *less* likelihood for food sharing and related cooperative behaviors than in times of low population, according to the model.

In Table 4 we examine our predictions for this effect. What we find is the opposite of what we expect. The marked cells indicate periods for which the model and the data disagree the most. Large positive residuals (indicating more cooperative behavior in the record than predicted; outlined in Table 4) are found in the context of relatively high populations. On the other hand, the negative residuals (indicating less cooperation than predicted; stippled in Table 4) are found mostly in the context of relatively low population levels. Unless we assume that the decisionmaking model, or the data, are seriously flawed, high population levels are promoting the desirability of cooperation in some way that more than counteracts the negative bias against cooperation that they should entrain by their effects on per-household income.

DISCUSSION

There are several ways in which the apparently strong effect of population size on promoting cooperation could be explained. One contributing factor might be that during this time population never reached a level in which per-household income was severely affected. This possibility is difficult to reject, although it has been possible to demonstrate that in particular situations, households have certainly had to work harder to maintain an acceptable level of income in times of high population and aggregation. For example, Kohler et al. (1986) were able to show that in the Dolores area, average (per household) distance to fields must have increased from only about 0.2 km., under conditions of low population and dispersion, to over 1.6 km., when

TABLE 4 Differences between predictions of risk-sensitivity model and archaeological data, examined against rough population index. Shaded cells indicate much less cooperative behavior than predicted; outlined cells indicate much more than predicted.

Period (A.D.)	Difference between predicted and observed rank (positive values indicate more cooperative behavior than predicted)	Ordinal index of probable regional population density (1 = lowest)[1]
1006–1029	-2	1
1030–1048	-1.5	1
1049–1088	0	2
1089–1099	-2.5	1
1100–1129	0	2
1130–1179	-.5	1
1180–1211	-1	2
1212–1221	3	2
1222–1271	3	3
1272–1288	1.5	1

[1] These are impressionistic. Relatively accurate population estimates can be made only for specific portions of the project area (especially the Dolores Archaeological Project area, Mockingbird Mesa, Upper Sand Canyon, and Lower Sand Canyon). These estimates, tabulated in Van West and Kohler (1995:Table 1) show that, generally, population increases and decreases are not synchronized within the study area, further complicating the task of making regional estimates.

population size was high and people lived in large villages (see also Orcutt et al. 1990).

However, even if per-household production was not diminished when populations levels were high, this by itself cannot explain what appears to be the *positive* effect that high populations have on cooperative behaviors (at best it would explain the absence of a negative effect). Where else can we look for some cause?

One place is in the nature of the landscape itself, with its rich array of natural and social features. The effects of large, sedentary populations of farmers on the biotic landscape are now beginning to be understood (Floyd and Kohler 1990; Kohler 1992b; Kohler and Matthews [1988]; Speth and Scott 1989). Wild-food depletion in conjunction with high population levels increased Anasazi commitment to and dependence upon agriculture. Under such conditions the landscape would have begun to fill up with marks of ownership (Adler 1990; Kohler 1992a), particularly where agriculture was most reliable and productive. In periods of both low population and low production, defection into an open landscape retaining abundant wild resources was more attractive than in periods of low production coupled with high population. We suggest that population levels—because of their effect on the landscape and on the required level of agricultural intensification—in effect changed the shape of the utility function to make sharing more attractive when population was high, and less attractive when population was low. From a game-theoretic perspective, the value of defection relative to cooperation was lower in populous landscapes. The dilemma of the 1270s/1280s in the study area was that the biotic and social characteristics of the landscape necessitated a sharing adaptation at the same time as climatic characteristics were making such cooperative behaviors increasingly unattractive. The abandonment of the Four Corners area may be due more to the undesirability of dispersing into a landscape that had lost its attractiveness for the dispersion/disintensification option, than to the absolute production levels that could have been achieved.

Our simple model predicts that cooperative behaviors such as aggregation can develop in the absence of marked increase in population, and at any population level. This prediction may in fact agree with the circumstances of aggregation in the Zuni (Stone 1992) and Taos (Crown et al. 1990) areas. If our interpretation of the weaknesses of the simple model is correct, however, the productive characteristics that favor pooling would have to be correspondingly more exaggerated to have the same effect in a sparsely populated region as in a densely populated region (see a parallel argument in Plog et al. 1988). The population history and level of the region are key contexts within which the utility model that influences decision-making must operate.

A second important area in which this simple model is almost certainly too simple is in its ignorance of extraregional productive characteristics. Presumably at least part of the population history of our study area can be explained by its attractiveness relative to other regions, in the same way as some population movements within the study area have been explained (Schlanger 1988). If so, this in turn will affect the shape of the utility function (or, in game-theoretic terms, the nature of the payoff matrix for cooperating versus defecting) within the study area.

Finally, to lop off some portion of the Southwest from the whole for analysis is to assume that developments in each subregion were independent. Clearly our study area is influenced by the development of the Chacoan system outside of its borders, and probably by other external events and systems that are not so obvious. This study area is one of convenience, justifiable mostly on the basis of the high-quality data that makes it possible to formulate and test, in very preliminary fashion, the present model. Nevertheless, our general success suggests that the formation and dispersion of villages in this time and place is largely conditioned by local factors, although the form and particularities of villages and their attendant social lives are undoubtedly affected by complex, local and nonlocal, considerations.

SUMMARY AND IMPLICATIONS

In this chapter we work toward a method for predicting the development of cooperative behaviorsfood, sharing among small-scale horticulturalists. The model behind the method focuses on sharing of food. Food sharing is not only one of the most fundamental forms of human cooperation but is symbolically implicated in more elaborate cooperative endeavors and may form the foundation for their elaboration. Our goal in this chapter has not been to define the links between food sharing and the various cooperative behaviors that leave more visible residues in the archaeological record. Nevertheless, examination of the performance of the model suggests that its utility goes well beyond the prediction of food sharing to the prediction of a large set of cooperative behaviors.

The model put forward here predicts that the effect of variance (risk) in production on cooperative systems involving food sharing will depend on whether these risks are situated in a context of relatively high, or relatively low, production. Cooperative behaviors are most valuable in the context of high production coupled with high temporal and spatial variability. Such behaviors are least likely in circumstances of low production coupled with high spatial and temporal variability. Essentially, the present method requires knowing something about the shape of the utility function before any predictions can be made about the relative value of cooperation versus defection. We have assumed that years of production substantially below the mean are low enough so that they will be located in the increasing-marginal-value segment of the curve, and that years well above the mean are within the decreasing-marginal-value segment of the curve. These assumptions require continued scrutiny. Provisionally, the success of the predictions they allow tends to confirm their general accuracy.

It has been realized by others that villages and expansive regional systems in the later northern Southwest tend edto form during periods that were favorable to agricultural production (see, for example, Orcutt et al. 1990; Sebastian 1991). In this chapter we build theory to explain *why* this should happen; this theory also successfully predicts when such cooperative systems should fall apart. One implication of this approach is that to turn climatic variability into a variable that is effective in human affairs, we first have to pass this variability through two filters—the first connecting climatic variability with production of economically significant materials (which we do with the retrodicted Palmer Drought Severity Indices and their resultant maize yields), the second connecting economy to cognition (which we attempt with the utility function).

The most important insight we take from this analysis is the surprising realization that a model of behavior based entirely on the self-interest of the participants—a rational choice model—can explain so much of the variability we see in the settlement record of this portion of the Mesa Verde region between A.D. 900 and 1300. Villages with their attendant public architecture form more or less when the model predicts they should, and disintegrate when they should, if the chief guiding tenet were the narrow economic self-interest of the participants. The success of methodological individualism (householdism?) in this context flies in the face of the received model of Puebloan adaptations as being highly group-oriented, submerging or effacing the individual.[9] This received model is, of course, based on historic groups, and it may well be correct for historic groups; these results minimally suggest some limits on backwards extrapolation from the "ethnographic present."

If so, it remains a fascinating empirical problem to determine at what point in prehistory or history the strong group-level effects featured in the received model became applicable to Puebloan societies. Certainly the ebb and flow of village life prior to about A.D. 1275, in comparison with the apparently uninterrupted history of village life after that time, suggest that an important threshold was reached about then. Perhaps with the many new techniques for water harvesting used in the northern Rio Grande, for example (Anschuetz 1992; Moore 1992), the productive conditions after A.D. 1275 always favored cooperation under a rational choice model: this is an important possibility that should be tested using approaches similar to those developed here for the Mesa Verde area. More likely, we think, is that other ways around the Prisoner's Dilemma were being exploited by this time. One

[9]These results also weaken any model that calls for the amassing of surplus by emergent leaders as *prerequisites* for village formation and the related cooperative behaviors we track in Table 2. By the model presented here, such general behaviors (although not, perhaps, their specific forms) can be explained more simply by households acting in their own best interests. Hantman (1989:442)—working on the same problem in the Upper Little Colorado Region—finds no evidence for storage volumes large enough to suggest that "economic surplus was produced and managed prior to the Pueblo IV period" of maximum local aggregation (and even then, may not be large enough to constitute a real "fund of power"). Our result also parallels recent work among insects (reviewed by Deneubourg and Goss 1989; see also Théraulaz and Deneubourg 1992) that demonstrate very complex social behaviors emerging from individuals following a few very simple rules in specific environments, without requiring direction by a dominant individual. We recognize, however, that once established, human villages—particularly as they exist in times of probable surplus—may present fertile environments for the emergence of dominant individuals or groups.

possibility is that the "shadow of the future" loomed larger in late prehistoric times; indeed, in retrospect we know that the probability of future interaction among villagers changed after 1275, but could the founders of an Early Classic village in the Rio Grande have been aware that their villages would endure longer, in some form, than were the inhabitants of a site such as Sand Canyon? Probably not. More likely, what changed was the open commitment of people to a new set of sharing rules that crosscut kin groups and featured village-level activities more strongly than had been the case in the past. Such commitments would have been accompanied by a stronger set of sanctions for potential defectors. Open commitments to a strategy and social sanctions against defection represent powerful methods for by-passing the Prisoner's Dilemma that had, up until about A.D. 1275/1300, dictated that villages endured only while times were good, and while it was in the best economic interest of all participants to continue to cooperate. Presumably, the katsina cult, the koshare system, and the remapping of kivas and social groups that all appear around this time provided both an attractive and a stable milieu for village life (Adams 1989; Lipe 1989; Steward 1937).

A related but more general way of viewing this transition draws on theory of human action developed by Elster (1989). According to Elster, people decide on various strategies for action—from among those that are practically available to them, within their "opportunity set"—according to either rational choice, or by reference to social norms. The present analysis suggests that rational choice was a dominant force in Anasazi settlement strategies until the late thirteenth century A.D., but that after that time, social norms came to dominate rational choice. If so, then this transition offers a remarkable opportunity for understanding how social norms may develop and take over decision areas that were formerly the domain of individual rational choice.

In closing, our approach suggests that disintegration of social groups beyond the level of close kin was a frequent way to cope with resource stress among the northern Anasazi prior to around A.D. 1275. Our approach also suggests that studies of sharing rules, and how they may change through time, is an area for critical future research. A final important implication of this study is that since the early villages and regional systems we have discussed are built on the backs of surpluses, we must become more sophisticated in our understanding of how these surpluses are mobilized and distributed (Saitta and Keene 1990); this problem is therefore closely connected with the problem of identifying sharing rules.

ACKNOWLEDGMENTS

This chapter has benefited from comments on related efforts by Linda Cordell, Richard Duncan, Michelle Hegmon, Keith Kintigh, John H. Miller, Rolf Sinclair, Eric Smith, and Chip Wills. We hope they forgive us for any good suggestions we chose not to follow. We would also like to thank Mark Varien and Angela Schwab of Crow Canyon Archaeological Center for access to unpublished data and manuscripts, and Larry Hammack and Jim Kleidon for recent tree-ring data from Knobby Knee Stockade and Sand Canyon Pueblo. All contributed information, reported more fully in Van West and Kohler (1995), helped make possible the test of the models presented here. William D. Lipe of Washington State University and Crow Canyon improved the clarity of our thinking and approach at several junctures, as did Jeffrey S. Dean of the Laboratory of Tree-Ring Research, University of Arizona. The Wenner-Gren Foundation for Anthropological Research through grant 4799 helped support the research by Van West that produced the annual estimates of prehistoric agricultural productivity on which the present analysis is based. The Graduate School of Washington State University made possible Kohler's professional leave, held at the Santa Fe Institute, providing Kohler with a stimulating environment while revising this chapter for publication.

REFERENCES

Adams, E. C.
1989 Changing Form and Function in Western Pueblo Ceremonial Architecture from A.D. 1000 to A.D. 1500. In *The Architecture of Social Integration in Prehistoric Pueblos*, edited by W. D. Lipe and M. Hegmon, pp. 155–160. Crow Canyon Archaeological Center, Cortez.

Adler, M. A.
1990 Population Aggregation and the Anasazi Social Landscape: The View from the Four Corners. Paper presented at the 1990 Southwest Symposium, January, Albuquerque.

Anschuetz, K. F.
1992 Saving a Rainy Day: The Integration of Diverse Agricultural Technologies to Harvest and Conserve Water in the Lower Rio Chama Valley, New Mexico. Paper presented at the New Mexico Archaeological Council Agricultural Symposium, October, Santa Fe.

Axelrod, R.
1984 *The Evolution of Cooperation.* Basic Books, New York.

Braun, D. P., and S. Plog
1982 Evolution of "Tribal" Social Networks: Theory and Prehistoric North American Evidence. *American Antiquity* 47:504–525.

Caraco, T., S. Martindale, and T. S. Whittam
1980 An Empirical Demonstration of Risk-Sensitive Foraging Preferences. *Animal Behavior* 28:820–830.

Crown, P. L., J. D. Orcutt, and T. A. Kohler
1990 Pueblo Cultures in Transition: The Northern Rio Grande. Revised version of paper presented at the conference "Pueblo Cultures in Transition: A.D. 1150–1350 in the American Southwest." Ms. on file, Crow Canyon Archaeological Center, Cortez.

Deneubourg, J. L., and S. Goss
1989 Collective Patterns and Decision-Making. *Ethology Ecology & Evolution* 1:295–311.

Elster, J.
1989 *Nuts and Bolts for the Social Sciences.* Cambridge, MA: Cambridge University Press.

Floyd, M. L., and T. A. Kohler
1990 Current Productivity and Prehistoric Use of Piñon (*Pinus edulis*, Pinaceae) in the Dolores Project Area, Southwestern Colorado. *Economic Botany* 44(2):141–156.

Ford, R. I.
1972 An Ecological Perspective on the Eastern Pueblos. In *New Perspectives on the Pueblos*, edited by A. Ortiz, pp. 1–17. University of New Mexico Press, Albuquerque.

Glassow, M. A.
1977 Population Aggregation and Systemic Change: Examples from the American Southwest. In *Explanation of Prehistoric Change*, edited by J. N. Hill, pp. 185–230. University of New Mexico Press, Albuquerque.

Gumerman, G. J., and J. S. Dean
1989 Prehistoric Cooperation and Competition in the Western Anasazi Area. In *Dynamics of Southwestern Prehistory*, edited by L. S. Cordell and G. J. Gumerman, pp. 99–148. Smithsonian Institution Press, Washington, D.C.

Haase, W. R.
1985 Domestic Water Conservation Among the Northern San Juan Anasazi. *Southwestern Lore* 51(2):15–27.

Hantman, J. L.
1989 Surplus Production and Complexity in the Upper Little Colorado Province, East-Central Arizona. In *The Sociopolitical Structure of Prehistoric Southwestern Societies*, edited by S. Upham, K. G. Lightfoot, and R. A. Jewett, pp. 363–388. Westview Press, Boulder.

Hegmon, M.
1989 Risk Reduction and Variation in Agricultural Economies: A Computer Simulation of Hopi Agriculture. *Research in Economic Anthropology* 11:89–121.

Kaplan, H., K. Hill, and A. M. Hurtado
1990 Risk, Foraging, and Food Sharing Among the Ache. In *Risk and Uncertainty in Tribal and Peasant Economies*, edited by E. H. Cashdan, pp. 107–143. Westview Press, Boulder.

Kohler, T. A.
1992a Fieldhouses, Villages, and the Tragedy of the Commons in the Early Northern Anasazi Southwest. *American Antiquity* 57:617–635.
1992b Prehistoric Human Impact on the Environment in the Upland North American Southwest. *Population & Environment* 13:255–268.

Kohler, T. A., and M. H. Matthews
1988 Long-Term Anasazi Land Use and Forest Reduction: A Case Study from Southwest Colorado. *American Antiquity* 53:537–564.

Kohler, T. A., J. D. Orcutt, E. Blinman, and K. L. Petersen
1986 Anasazi Spreadsheets: The Cost of Doing Agricultural Business in Prehistoric Dolores. In *Dolores Archaeological Program: Final Synthetic Report*, compiled by D. A. Breternitz, C. K. Robinson, and G. T. Gross, pp. 525–538. USDI, Bureau of Reclamation, Denver.

Krebs, J. R., and A. Kacelnik
1991 Decision Making. In *Behavioral Ecology: An Evolutionary Approach.* Third edition, edited by J. R. Krebs and N. B. Davies, pp. 105–136. Blackwell Scientific Publications, Oxford.

Leonard, R. A., and H. Reed
1993 Population Aggrgegation in the Prehistoric American Southwest: A Selectionist Perspective. *American Antiquity* 58:648–661.

Lipe, W. D.
1989 Social Scale of Mesa Verde Anasazi Kivas. In *The Architecture of Social Integration in Prehistoric Pueblos*, edited by W. D. Lipe and M. Hegmon, pp. 53–71. Crow Canyon Archaeological Center, Cortez.

Moore, J. L.
1992 Anasazi Field Systems in the Taos District. Paper presented at the New Mexico Archaeological Council Agricultural Symposium, October, Santa Fe.

Orcutt, J. D., E. Blinman, and T. A. Kohler
1990 Explanation of Population Aggregation in the Mesa Verde Region Prior to A.D. 900. In *Perspectives on Southwestern History*, edited by P. E. Minnis and C. L. Redman, pp. 196–212. Westview Press, Boulder.

Plog, F. T.
1974 *The Study of Prehistoric Change.* Academic Press, New York.

Plog, F., G. J. Gumerman, R. C. Euler, J. S. Dean, R. H. Hevly, and T. V. N. Karlstrom
1988 Anasazi Adaptive Strategies: The Model, Predictions, and Results. In *The Anasazi in a Changing Environment*, edited by G. J. Gumerman, pp. 230–276. Cambridge University Press, Cambridge.

Rima, I. H.
1972 *Development of Economic Analysis.* Richard D. Irwin, Homewood, IL.

Saitta, D. J., and A. S. Keene
1990 Politics and Surplus Flow in Prehistoric Communal Societies. In *The Evolution of Political Systems: Sociopolitics in Small-Scale Sedentary Societies*, edited by Steadman Upham, pp. 203–224. School of American Research, Santa Fe, and Cambridge University Press, Cambridge.

Schlanger, S. H.
1988 Patterns of Population Movement and Long-Term Population Growth in Southwestern Colorado. *American Antiquity* 53:773–793.

Sebastian, L.
1991 Sociopolitical Complexity and the Chaco System. In *Chaco and Hohokam: Prehistoric Regional Systems in the American Southwest*, edited by P. L. Crown and W. J. Judge, pp. 109–134. School of American Research Press, Santa Fe.

Sigmund, K.
1993 *Games of Life: Explorations in Ecology, Evolution, and Behaviour.* Oxford University Press, Oxford.

Smith, E. A.
1988 Risk and Uncertainty in the "Original Affluent Society": Evolutionary Ecology of Resource-Sharing and Land Tenure. In *Hunters and Gatherers 1: History, Evolution, and Change*, edited by T. Ingold, D. Riches, and J. Woodburn, pp. 222–251. Berg, Oxford.

Smith, E. A., and R. Boyd.
1990 Risk and Reciprocity: Hunter-Gatherer Socioecology and the Problem of Collective Action. In *Risk and Uncertainty in Tribal and Peasant Economies*, edited by E. Cashdan, pp. 167–191. Westview Press, Boulder.

Speth, J. D., and S. L. Scott
1989 Horticulture and Large-Mammal Hunting: The Role of Resource Depletion and the Constraints of Time and Labor. In *Farmers as Hunters: The Implications of Sedentism*, edited by S. Kent, pp. 71–79. Cambridge University Press, Cambridge.

Stephens, D. W.
1990 Risk and Incomplete Information in Behavioral Ecology. In *Risk and Uncertainty in Tribal and Peasant Economies*, edited by Elizabeth Cashdan, pp. 19–46. Westview Press, Boulder.

Steward, J.
1937 Ecological Aspects of Southwestern Society. *Anthropos* 32:87–104.

Stone, T. T.
1992 *The Process of Aggregation in the American Southwest: A Case Study from Zuni, New Mexico.* Unpublished Ph.D. dissertation, Department of Anthropology, Arizona State University, Tempe.

Théraulaz, G., and J.-L. Deneubourg
1992 Swarm Intelligence in Social Insects and the Emergence of Cultural Swarm Patterns. In *The Ethological Roots of Culture*, edited by A. Gardner. NATO ASI Series D.

Upham, S.
1984 Ecological and Political Perspectives on Labor Intensive Agriculture and Exchange During the 14th Century. In *Prehistoric Agricultural Strategies in the Southwest*, edited by S. K. and P. R. Fish, pp. 291–307. Anthropological Research Papers 33. Arizona State University, Tempe.

Van West, C. R.
1994 *Modeling Prehistoric Agricultural Productivity in Southwestern Colorado: A GIS Approach.* Reports of Investigations 67. Department of Anthropology, Washington State University, Pullman.

Van West, C. R., and T. A. Kohler
1994 A Time to Rend, A Time to Sew: New Perspectives on Northern Anasazi Sociopolitical Development in Later Prehistory. In *Anthropology, Space, and Geographic Information Systems*, edited by M. Aldenderfer and H. D. G. Maschner. Oxford University Press, New York.

White, T. D.
1992 *Prehistoric Cannibalism at Mancos 5MTUMR-2346.* Princeton University Press, Princeton.

Winterhalder, B.
1986 Diet Choice, Risk, and Food Sharing in a Stochastic Environment. *Journal Anthropological Archaeology* 5:369–392.
1990 Open Field, Common Pot: Harvest Variability and Risk Avoidance in Agricultural and Foraging Societies. In *Risk and Uncertainty in Tribal and Peasant Economies*, edited by E. Cashdan, pp. 67–87. Westview Press, Boulder.

Alison E. Rautman
Department of Anthropology, Michigan State University, East Lansing MI 48824

Risk, Reciprocity, and the Operation of Social Networks

For many archaeologists, deciphering cultural practices in the prehistoric Southwest and investigating the reasons behind observed changes in cultural behavior are problems inseparable from an understanding of the climatic fluctuations that form part of the environmental context of prehistoric societies in this region (e.g., Dean et al. 1985; Cordell and Gumerman 1989). There are many problems, however, with attempting such an analysis of the relationship between climate and human behavior. First, there are so many different variables associated with climate and environment that, as Plog and Hantman (1990) point out, it is not that difficult to find some type of environmental change that is roughly contemporaneous with a given sociocultural event. Second, there has to be some way to interpret the importance and meaning of such correlations for societies in the past.

Some archaeologists have taken a methodological approach to this problem, investigating to what extent our (archaeological) interpretations have presented us with spurious temporal correlations between human behavior and climatic factors. Plog and Hantman (1990), for example, suggest that the contrast in temporal scale between the measurements of climatic change and of coarser-scaled cultural phases has made it difficult to sort out the actual temporal relationships between climatic events and cultural responses. They suggest that this situation can be remedied in part by improving our determination of chronologies so that we will be better able to detect smaller-scale patterning and correlations among cultural variables, and between cultural and climatic variables.

However, the problem is not only one of detecting correlations, and rejecting spurious ones, but also of understanding how climate affects human actions. Humans react to their environment, and in turn affect their environment, through an intermediary: the sociocultural system. Social variables such as demography (including population size and distribution), and technology and technological organization (Cordell and Plog 1979), affect what climatic variables will be important to human groups, and affect how each may structure peoples' options and opportunities for action and reaction.

Accordingly, archaeologists and anthropologists studying the interrelations between environment and culture have also become interested in understanding the sociocultural factors that affect perceptions of climate, environment, and definition of resources (e.g., Cronen 1983). Others (e.g., contributors in Cashdan 1990) focus on the cultural definition of risk, and responses to risk in different sociocultural contexts. In doing so, there is implicit acknowledgement that other cultural perceptions of identical environmental conditions are possible under different circumstances, that the effects of climatic and culture change may operate at a variety of temporal and spatial scales, and that the patterning that we perceive in the archaeological record need not represent strategies that are necessarily stable or effective in the long term (e.g., Ortiz 1990).

Defining what is meant by risk, or identifying important levels of environmental variability, therefore, is not a topic that can be addressed by the anthropologist or archaeologist without reference to the social group that would have perceived it or reacted to it. For example, archaeologists can identify patterns in the climatic record, such as periods of lower-than-average precipitation, but evaluating the importance of these patterns to a prehistoric society represents yet another level of modeling.

This chapter describes a first attempt at such a level of modeling by examining how social networks might have operated to reduce risk by providing a context for reciprocal exchanges among groups of hunter-gatherers or part-time agriculturists in prehistoric central New Mexico. A social network represents numerous dyadic interactions among individuals or groups of people, and can be described by a large number of variables, including its "size" or geographic location of its members, and its "shape" or geographic directionality. In addition, network operation might include variables that measure, for example, the intensity of social interaction among network members, the nature and timing of their interactions, the duration of network relationships, the continuity of interactions among network participants, and so forth.

This study begins by evaluating the expected size (spatial extent) and shape (geographic directionality) of social networks in prehistoric central New Mexico. Modern climatic information is used to evaluate the extant spatial and temporal patterning of environmental variability in this region, and to predict the expected minimum size and preferred geographical areas within which social interactions could facilitate information transmission, group mobility, or resource exchange that could reduce the risk of subsistence stress associated with climatic variability within the region. The potential for reciprocal relationships among egalitarian societies within such a network is then considered. Network operations are not considered for the entire region, but rather from the perspective of small-scale agriculturists who occupied one particular site called the Kite Site Pithouse Village (LA-38448) from about A.D. 980 to 1250.

COPING WITH RISK: THE ROLE OF SOCIAL NETWORKS

Among nonhierarchical societies such as that which occupied the Kite Site during the Pithouse Period, a number of social and cultural means exist to assess and respond to environmental variability. Both Nelson (1995) and Minnis (1985, 1995) review some of the variety of behavioral and technological responses that have been observed among ethnographic societies.

This study focuses particularly on the strategy of participation in a regional social network by individual households and local groups. Households have many different ways of reducing different kinds of risk (e.g., Hegmon 1995); participation in regional social networks is expected to be effective in mitigating only a certain type of risk, particularly that which is engendered by differential spatial distribution of resources. The social interactions within the network provide a way for people to exchange information about conditions in different areas, exchange resources from different areas, and/or define areas where people in need could go to exploit resources (Braun and Plog 1982; see also Renfrew 1975).

The role of social networks in buffering subsistence risk that is caused by variability in the natural environment stems from the fact that networks have a spatial component: network members reduce their risk of suffering stress by ensuring their own access to resources of another area. Dean et al. (1985:543) can state, therefore, that the utility of interaction and exchange in coping with local subsistence stress will depend on the spatial patterning of environmental variability. Plog (1983) has explicitly investigated this spatial component of variability in structuring interactions among groups in different geographic regions.

The roles that social networks play in a given society, however, are not limited to buffering risk, nor are they necessarily static through time. Upham (1982) and Plog (1983, 1984), for example, have suggested that in some cases, the changing character of alliances between local networks can effect organizational change that leads to the development of central sites, specialized production, and possibly social ranking and stratification.

The importance of social networks, and the social interactions that occur within them, stems from the fact that all local groups are already presumably participating in some form of extra-local interaction, if only to provide social contacts that are necessary to maintain a mating network of sufficient size over time (Wobst 1974). The existence of some sort of regional social network is therefore assumed; it is the structure and character of the social networks that is expected to differ among societies. Braun and Plog (1982:508) suggest, for example, that the "social connectedness" of these networks, or the degree of integration among network participants, will be responsive to the duration and amount ("level") of uncertainty and risk associated with regional environmental variability.

Regional social networks extending beyond the local group reduce the risk of resource stress by providing information about conditions in distant areas as well as potentially providing access to distant resources. Among hunter-gatherers and small-scale agriculturists, the actual amount of food exchanged within a social network may be minimal, but the social network provides a potential mechanism for identifying more productive areas, and then redistributing personnel to them (Colson 1979; Yengoyan 1972).

The social contexts of exchange and the nature of intergroup contact may also vary, depending on the social distance between groups. Among hunter-gatherers, social relationships are maintained by regular interaction and exchange of goods and/or labor within the context of institutionalized reciprocity (Bailey and Peacock 1988; Wiessner 1977). Institutions of feasting, exchanges of goods and labor, and gift giving between social units characterize middle-range and complex societies as well (Bailey and Peacock 1988; Ford 1972; Scudder 1962). In some societies, ritual obligations mediate the flow of material goods and structure interaction; ritual activities can also regulate and enforce interactions among socially distant groups, ensuring network continuity even during times when its role in risk reduction may not be particularly significant (Bean 1972; Ford 1972; Minc 1986; O'Shea 1981).

If it is true that social networks can and do function to reduce risk, then two research questions present themselves. First, how is risk to be defined? In a given situation, what aspects of environmental variability contribute most to the risk that can be buffered by regional social interactions?

Second, if we can identify selected dimensions of risk using independent (noncultural) evidence, it should be possible to predict (or retrodict) some aspects of network operation: for example, the expected minimum size of social networks that would be effective in reducing this risk.

It is assumed here that cultural response to risk will match, in capacity and scale, that variability with which it is intended to cope (Clarke 1985; Minnis 1985). The structure of environmental variability is therefore particularly important for analysis because it provides an independent predictor of the structure of effective cultural response (Halstead and O'Shea 1989). While there is little direct relationship between climate and human behavior, a study of environmental variability can provide an estimate of the relative vulnerability of certain areas to unpredictable and stressful conditions, and monitor the synchrony of conditions among areas. It can thereby provide a means of evaluating the potential effectiveness of cultural buffering strategies, such as regional social interactions, in coping with such conditions.

In other words, it is expected that not all loci in a region will be equally vulnerable to certain types of environmental variability, and thus not all possible loci will be equally valued as nodes in a potential social network. Alternative resource areas are expected to be those experiencing relatively low vulnerability to specific stresses engendered by environmental variability, particularly for times during which a selected area might be experiencing lowered productivity and resource stress.

It is also expected that the intensity of interaction and exchange will not decline equally in all directions from a point source (e.g., Hantman and Plog 1982). The constraints on interaction posed by the relative vulnerability of alternate areas to environmental variability focuses attention on maintaining access to resources in some areas at the expense of others. In other words, the geographic "shape" of social networks includes a directional component, which indicates differential intensity or frequency of interaction with groups in preferred areas.

In this study, the process of investigating the structure of variability in the environment is begun by examining climatic factors. Evaluating the relative vulnerability of areas to impact from climatic variation allows definition of feasible alternate resource areas for use during times of local subsistence stress caused by extended periods of high variability, or by unusually poor conditions. In this study, climatic information provides an independent estimator of the size and directionality of an expected social network that would be necessary for inhabitants of a particular area to mitigate local resource variability.

The definition of the environment itself depends on the scale of analysis, and adaptive responses do not take place in a cultural vacuum, responsive only to climate. The history of the system also forms part of the system's environment (Winterhalder 1980), while cultural factors such as demography, the organization of production, and the structure of social relations may alter the effects of a given type of variability on the cultural system, as well as the possible structure and effectiveness of a given buffering strategy (Cordell and Plog 1979; O'Shea and Halstead 1989).

Concern with the environmental context of decision making therefore does not imply a deterministic framework, since the conceptual focus of analysis is the interaction between humans and their environment. The environment forms part of the context for decision making among a variety of alternative actions; actual selection and use of a given strategy, such as participation in a social network, depends also on the social and cultural context. However, to be effective, the strategy selected must act to reduce that aspect of risk toward which it is directed (e.g., Minnis 1985). It is in this general sense that predictions of effective cultural responses can be made.

CLIMATIC VARIABILITY IN CENTRAL NEW MEXICO

In the American Southwest, an excellent dendroclimatic record has provided a great deal of information that can be used to reconstruct paleoclimatic patterning and its spatial and temporal variability (Dean 1995; Dean and Robinson 1977, 1978). These reconstructions have been extensively used by archaeologists, particularly to model large-scale patterns of social interaction between geographic regions (e.g., Plog 1983). For this study, however, these reconstructions encompass too large a scale; the data points are too widespread for investigating within-region interactions. Use of modern climatic data (U.S. Department of Commerce 1917–1980) therefore proved to be necessary to develop a model of the structure of climatic variability at the desired spatial scale in this region.

Modeling the expected spatial extent and directionality of prehistoric social networks with modern climatic data is obviously problematic if the modern climatic data are simply projected into the past. Here, the modern numerical values for climatic variables are not intended to be considered culturally significant, nor are the actual values for the climatic variables intended to represent conditions during the prehistoric period.

Rather, modern climatic conditions are used as proxy indicators of the spatial patterning of selected aspects of variability; thus, analysis focuses on identifying spatial and temporal correlations, or the degree of synchrony, of conditions throughout the study area. The most important (and of course debatable) assumption is that this overall structure of environmental variability is relatively stable over time, although of course the correlations among stations may diverge considerably from the overall pattern during any given year.

A culturally meaningful study area was defined by reference to ethnographic studies of land use and mobility among egalitarian societies. Since the prehistoric occupants at the Kite Site are known to have had at least a part-agricultural economy (Rautman 1990), data from contemporary hunter-gatherers in a semiarid environment provides a conservative and generous estimate of the size of region that might be occupied by a breeding population in generally similar regions.

Definition of the geographic size of the study area follows Plog and Powell's (1984) procedure (described in Rautman 1990). According to these criteria, an area about 20,000 $km.^2$ in size was defined by constructing an imaginary circle 80 km. in radius around the Kite Site. Modern weather stations included within this circle form point locations for monitoring climatic variables. Where possible, modern weather stations that were located outside the circle, but nearby, were also included (Figure 1). For simplicity, social networks will be conceived as dyadic relationships between inhabitants of the Kite Site and people located near the other weather station points. Conditions at the Kite Site are estimated by reference to the closest weather station, Gran Quivira (see Figure 1).

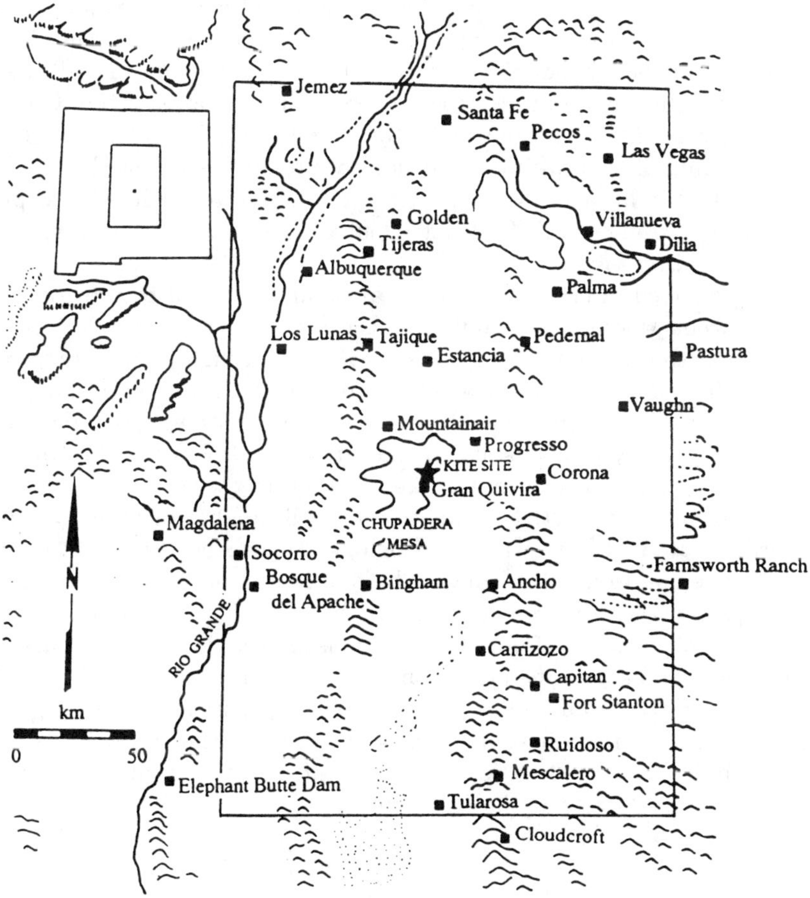

FIGURE 1 Location map showing the study area, major topographic features, and the location of modern weather stations used in the analysis (from Rautman 1993).

In this analysis, the spatial patterning of climatic variables is considered to be of particular importance. First, the spatial extent of climatic factors that affect the productivity and predictability of subsistence resources is monitored by examining the co-occurrence of specified conditions (such as precipitation levels) throughout the study area. This portion of the analysis attempts to identify the spatial extent of conditions occurring at the Kite Site.

When conditions at the Kite Site are poor, it is expected that the inhabitants would seek alternative sources of resources in other areas. "Poor" conditions are rather arbitrarily defined as years of below-average precipitation, and are presumed to represent at least some level of lowered productivity for local plant and animal resources. The validity of this assumption is addressed below. Years that are considered to be "poor" for resource exploitation at the Kite Site were thus identified, and conditions at other stations during these years were then evaluated to estimate their potential utility for resource procurement by inhabitants of the Kite Site.

In order for a station to rank highly as a preferred alternative resource area, several criteria must be met. First, conditions at preferred stations should generally not co-vary with conditions at the Kite Site. It is therefore expected that proposed alternative resource areas will be located outside the area of highly synchronous conditions defined above

Second, preferred alternative resource areas are intended to provide resources during years when resource availability at the Kite Site may be relatively low. It is thus expected that such stations would experience a high proportion of years of "good" conditions during years in which conditions at Gran Quivira are poor.

Third, because of the time and effort involved in maintaining social interactions at a distance, favorably ranked stations that are located closer to the Kite Site are expected to be preferred over those located further away.

These criteria are expected to identify a small number of feasible alternative resource areas, but single numerical ranking of stations is not expected. It is doubtful that any given station will be always be free of unusually poor conditions, and it is also expected that different criteria (such as a different definition of "poor" conditions) will result in a different ranking. It is expected, however, that a limited number of areas can be identified that commonly experience climatic conditions out of synchrony with the Gran Quivira area; the spatial distribution of these stations provides an estimate of the spatial extent and as well as the "shape" of social networks that would be needed to cope with the effects of environmental variability and possible resource stress at the Kite Site.

THE KITE SITE SOCIAL NETWORK

EXPECTED MINIMUM SIZE OF THE KITE SITE SOCIAL NETWORK

The spatial extent of conditions that affect the Kite Site area is evaluated by cluster analyses, using Ward's method on standardized z-scores. Thirty-seven cluster analyses were run, one for each complete year of record at Gran Quivira from 1940 to 1980. The resulting clusters group together weather stations that experienced similar conditions of rainfall amount, temporal distribution of rainfall, and frost-free season length. Member stations for each cluster solution were then examined to identify which stations cluster together with Gran Quivira each year over time.

This station ranking was then divided by inspection into six groups, based on the percentage of years that a given station covaries with Gran Quivira (Figure 2).

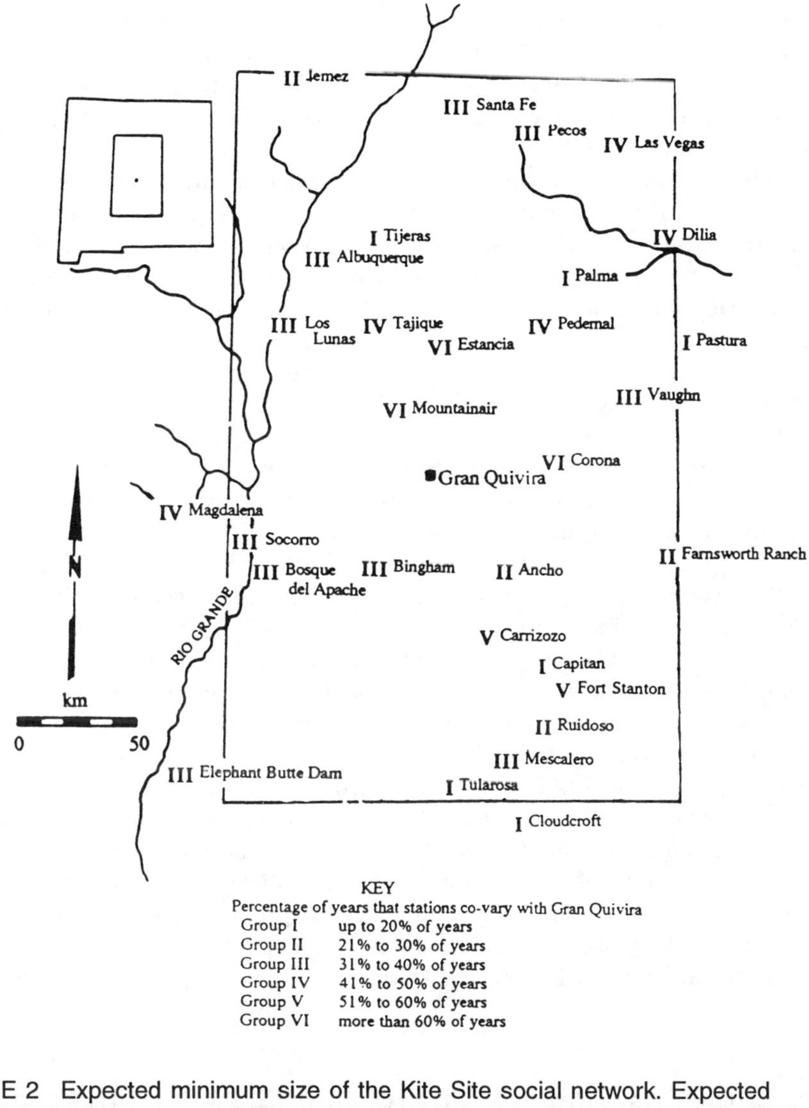

FIGURE 2 Expected minimum size of the Kite Site social network. Expected alternative resource areas would lie outside the area that commonly experiences conditions that are similar to those at the Kite Site. Group V and VI stations are therefore expected to mark low-priority alternative resource areas; Group I and II stations are high-priority areas (from Rautman 1993).

Group I includes those stations that cluster with Gran Quivira less than or equal to 20 percent of years; these stations include Capitan, Cloudcroft, Palma, Pastura, Tijeras Canyon, and Tularosa.

Group II includes those stations clustering with Gran Quivira from 21 to 30 percent of all years: Ancho, Farnsworth Ranch, Jemez, and Ruidoso are members of this group. Stations in Groups I and II are considered to be eligible as alternative resource areas, since they commonly experience climatic conditions that are different from those at Gran Quivira.

The other groups of stations experience conditions similar to those at Gran Quivira more than 30 percent of the time, and would probably be less desirable as alternative resource areas. Group III stations co-vary with Gran Quivira 31 to 40 percent of recorded years and include Albuquerque, Bingham, Bosque del Apache, Elephant Butte, Los Lunas, Mescalero, Pecos, Santa Fe, Socorro, and Vaughn.

Stations in Group IV covary with Gran Quivira from 41 to 50 percent of recorded years; these stations are Dilia, Las Vegas, Magdalena, Pedernal, and Tajique. Group V stations cluster with Gran Quivira 51 to 60 percent of years; these stations are Carrizozo and Fort Stanton. Group VI stations cluster together with the Gran Quivira station more than 60 percent of recorded years; this group includes only Corona, Estancia, and Mountainair.

These analyses show that the areas around Corona, Estancia, and Mountainair (Group VI stations) are experiencing conditions similar to those at the Kite Site more than 60 percent of the time. Climatic variables such as precipitation and frost that affect conditions at Gran Quivira can be assumed, therefore, to encompass the area that is broadly defined by the location of these four weather stations.

Of course, there are still many years (30 percent) when conditions at Estancia (for example) differ from those at the Kite Site. During these years, social relations with groups near Estancia may indeed have provided needed access to resources in that area. And of course the area near Estancia may have desirable resources (such as salt or stone) that would justify continued social interaction regardless of climatic conditions. This portion of the analysis, however, indicates that to the extent that social contacts with other groups are intended to provide resources during times of low productivity, groups at the Kite Site are expected to seek such contacts outside this "local area," within which stations commonly experience conditions in synchrony.

THE EXPECTED GEOGRAPHIC "SHAPE" OF THE SOCIAL NETWORK

The next step of the analysis investigates possible rankings of stations outside the local area. If all such stations were equally attractive to inhabitants of the Kite Site, the spatial pattern of social interactions in the study area might resemble a doughnut shape, with relatively few interactions within the "local area" and more numerous interactions surrounding it. However, not all social groups located outside this area are expected that to be equally "useful" to the inhabitants of the Kite Site for the purpose of buffering low productivity (although of course they may be "useful" for some other purpose).

Furthermore, if we consider that establishing and maintaining social interactions does involve at least some social effort, it is unlikely that the same kind of relationship will characterize interactions with every group in the region. It is expected groups in some areas will have to be comparatively neglected in favor of others. The social network is therefore expected to contain a directional component, with more interaction (or greater intensity of interaction) occurring in relatively few preferred directions.

Since this study addresses particularly the role of social networks in mitigating the risk of resource stress, the spatial patterning of years of low productivity is considered to be particularly significant in structuring social interaction within the region. Alternate areas for resource exploitation are assumed to be loci that are somehow "better."

The problem, then, is developing a useful definition of "poor" and "good" conditions that allow comparison among stations, given the spatial and temporal scale of the data available, and its characteristics. For example, Minnis (1985) defines stressful conditions in reference to the number of years that precipitation levels fall below one standard deviation below the mean. His definitions provided an excellent way of monitoring climatic conditions over a long time period in the Mimbres area, but proved to be less useful for working with short-term climatic information in central New Mexico. For example, according to these criteria, only 6 years at the Kite Site (14 percent of years in the available climatic record) could be considered to be "stressful" for local inhabitants—a fact that highlights the relatively favorable local conditions extant at the Kite Site, but provides an extremely small sample of "stress years" for regional comparisons.

An alternative working definition of "poor" years was therefore adopted—one that focuses on variation around the station mean. Ecologists assume that, very generally speaking, floral and faunal resources found at a given area are adapted to average conditions at that area (Wiens et al. 1986). "Poor" years were therefore defined as those during which total annual precipitation fell below the station mean; "good" years are those during which station conditions were at least average. Note that stations are evaluated only in reference to their own mean conditions, not the regional mean.

The status of each station as a feasible alternative resource area is estimated by the number of years that a station is selected as "good" during "poor years" at the Kite Site. For example, during 40 years of weather record at Gran Quivira, there are 23 years during which Gran Quivira weather station received precipitation measuring less than the mean amount for all years. These years are defined as "poor," and may represent some level of lowered local resource availability for populations in the area around the Kite Site.

A rather large imaginary chart might list those 23 poor years along the y-axis and all other weather stations along the x-axis. A mark in each box of the matrix so constructed can be used to indicate whether conditions at each station are good (equal to or above its station mean), or poor (below its station mean). The marks are then totaled by station (column) to determine the number of times that station is experiencing good conditions. This number is then divided by 23 to express the percentage of years for which that given station would be experiencing relatively good conditions during stress years at Gran Quivira.

All 34 stations used in the precipitation study experienced at least one good year while conditions at Gran Quivira were poor. Values range from a low of 4 percent (at Pastura and Mountainair) to a high of 39 percent (Tularosa).

These values are divided by inspection into four groups (Figure 3). Group A stations experience good conditions during 1 to 9 percent of stress years at Gran Quivira; Group B stations would be feasible alternatives during 10 to 19 percent of stress years; Group C stations experience at least average conditions during 20 to 29 percent of stress years at Gran Quivira. Group D stations experience at least average conditions during 30 to 40 percent of poor years at Gran Quivira.

Group D stations (Corona, Palma, and Tularosa) are considered to be the best alternative resource areas. Conditions at these stations are more commonly out of synchrony with conditions at Gran Quivira, experiencing good conditions during many years for which Gran Quivira is under stress.

It is especially interesting to note that the values for mean annual precipitation and for precipitation variance in these areas are not particularly favorable when these stations are compared with all others in the study area—Tularosa has very low mean annual precipitation, for example, and Corona and Palma experience relatively high variability in annual precipitation levels. However, when conditions are poor at Gran Quivira, these are the stations with the highest probability for good local conditions.

The groupings on this figure show that no one factor such as overall precipitation levels, precipitation predictability, general geographic setting, or elevation can be used to predict alternative resource areas that might be preferred for coping with low productivity at the Kite Site. Rather, I suggest that additional information regarding past and/or current conditions at a given area would have been required in order for human groups to be able to evaluate the relative utility of different loci for this purpose.

PREFERRED LOCATIONS FOR INTENSIVE NETWORK INTERACTIONS

If it is assumed that the distances involved in either movement of people or transport of goods is a factor in evaluating potential alternative resource areas, the closer area—that around Corona and Ancho, approximately 50 km. away—is predicted to be the highest priority area for obtaining alternative resources when conditions at the Kite Site are poor and food resources may be lower than usual (Figure 4). This result is particularly interesting because these areas did not seem to be especially favored in other respects. For example, conditions of precipitation and frost-free season in these two areas are quite similar to those at the Kite Site. In fact, in many respects these areas are generally *less* favorable: productivity in these areas is lower than the regional average, and variability in interannual productivity is high.

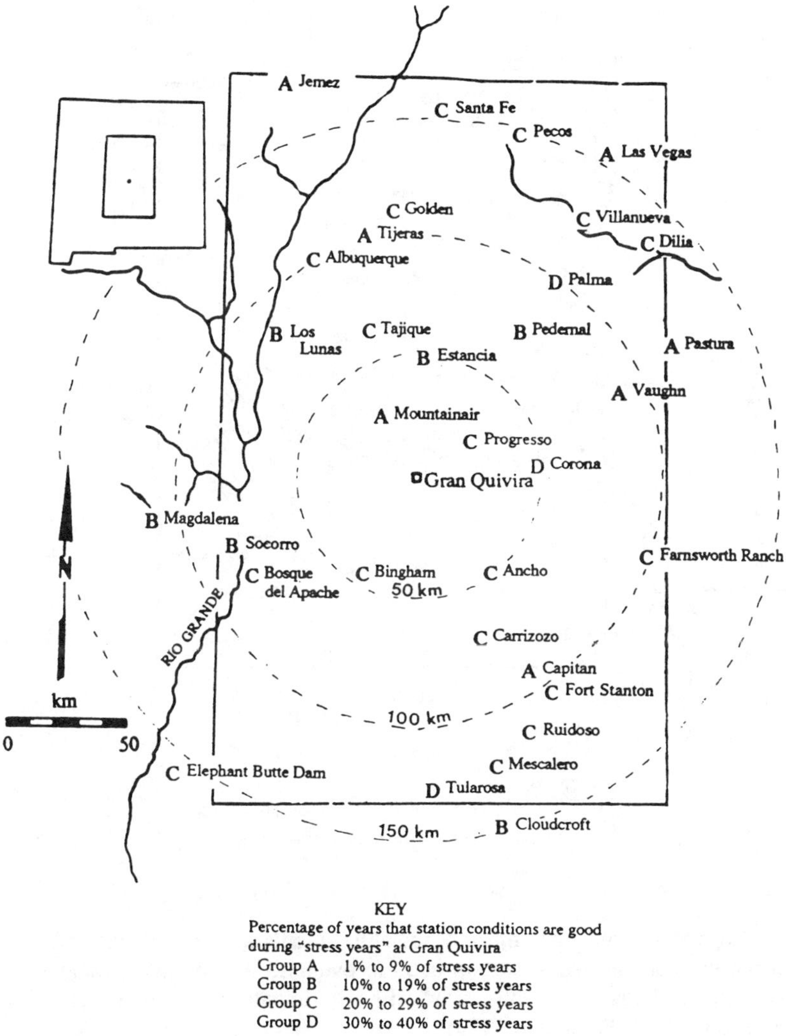

FIGURE 3 Expected "shape" of the Kite Site social network. Expected alternative resource areas are those that are most likely to be experiencing at least average conditions during "poor" years at the Kite Site, when there is a potential for local resource shortfalls. Group A stations would be the least likely to be selected as alternative resource areas; Group D stations would be the most likely to be selected (from Rautman 1993).

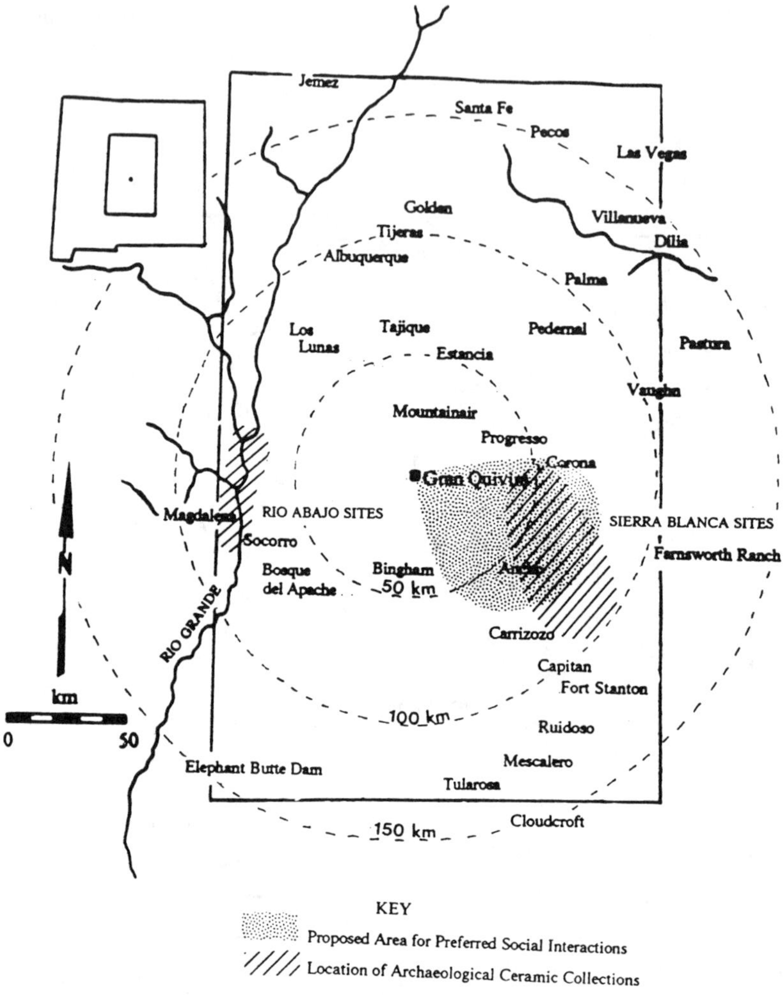

FIGURE 4 Proposed area for preferred social interactions (stippled pattern). The location of archaeological assemblages used to evaluate the model's predictions is marked by diagonal lines. Concentric dashed lines indicate the approximate distance from the Kite Site (from Rautman 1993).

However, when one considers the synchrony of climatic conditions, then these stations rank quite high. Ancho stands out because it is the closest station that commonly experiences conditions that are different from those at the Kite Site. Corona, however, has very nearly the same pattern of intra-annual variability as does the Kite Site, and this station is also surprisingly similar in overall climatic patterning as well (clustering of monthly precipitation and frost-free season length grouped Corona and Gran Quivira together for 65 percent of all years). The proposed importance of Corona for inhabitants of the Kite Site stems from the fact that it commonly experiences average or above-average conditions during poor years at the Kite Site—those 53 percent of years when precipitation at Gran Quivira is below average. At Corona, 30 percent of years are average or better during poor years at the Kite Site; this proportion, although low, is the highest observed among all weather stations in central New Mexico.

PREHISTORIC SOCIAL INTERACTIONS IN CENTRAL NEW MEXICO: AN EVALUATION OF THE MODEL

The nature of interactions between inhabitants of the Kite Site and people in other areas is not specified by the climatic model. Interactions could involve regular or episodic exchange of food, mates, and information, or might be more restricted. Similarly, resources in preferred alternative resource areas might be accessed by group mobility or by food exchange. For the purposes of evaluating the model, a measure of interaction is needed, but it is not necessary to test the nature of this interaction.

Ceramic styles and ceramic assemblage similarity have been used as an indicator of prehistoric social interaction (e.g., Braun and Plog 1982; Plog 1976). Similarity of ceramic styles may result from a number of processes, including actual exchange of painted vessels or some level of interaction that results in adoption of an extant style to decorate locally made vessels. Distinguishing between these possibilities is a problem of interest for many archaeologists, but is not at issue here. The model specifies only that interaction between the Kite Site and preferred areas should be greater than between the Kite Site and lower ranked areas.

It is therefore expected that ceramic assemblages at sites around Corona and Ancho should exhibit greater similarity to the Kite Site assemblage than do sites of comparable age in other areas that are located a comparable distance away. Ceramic assemblages are not expected to be identical, since social networks are maintained by groups in each area for a variety of reasons.

This expectation is tested by analysis of ceramic assemblages from archaeological sites around Corona and Ancho. This area is part of the Sierra Blanca region of the Jornada Mogollon (Kelley 1984); Corona and Glencoe Phase sites are considered to be approximately contemporaneous with the Pithouse Period occupation at the Kite Site. These two phases are both dated to approximately A.D. 1100 to 1200 (Kelley 1984:44–51). Corona Phase sites are found in the northern part of the Sierra Blanca region (near the modern town of Corona); contemporaneous Glencoe Phase sites are located further south in the same region (Kelley 1984:Figure 9).

In comparison, ceramic assemblages from sites of comparable date (and comparable distance away) in other areas are expected to be less similar to the Kite Site. The Rio Abajo region near the modern town of Socorro was selected to provide a contrast to the Kite Site-Sierra Blanca dyad. The Elmendorf Phase of the Rio Abajo region (Marshall and Walt 1984) is considered to be approximately contemporaneous with the pithouse occupation of the Kite Site. The Early Elmendorf Phase is dated to about A.D. 950 to 1100; the Late Elmendorf Phase is from A.D. 1100 to 1300 (Marshall and Walt 1984:75–95). Since the Kite Site would have been occupied throughout this time, sites of Early and Late Elmendorf were grouped together for this analysis.

Interaction between communities in different regions is evaluated by calculating the degree of similarity of ceramic assemblages (Plog 1976). The Brainerd-Robinson coefficient (Brainerd 1951; Robinson 1951) is used here; this coefficient measures similarity of assemblages by comparing the proportional representation of each category within each assemblage.

For this analysis, the percentage representation of each ceramic type in a region is compared with the representation of the same type in other regions; the absolute value of the difference in the percentage representation is noted. These differences in percentages are summed and the result is subtracted from 200; this figure is the value of the Brainerd-Robinson coefficient. The coefficient ranges in value from 0 to 200; a value of 200 represents maximum similarity between assemblages, and a value of zero represents maximum dissimilarity.

Brainerd-Robinson coefficients were calculated in three ways, each of which confirms the greater similarity of ceramic assemblages between the Kite Site and the Sierra Blanca region. Similarity of the general assemblage structure was investigated first for general ware categories. Additional calculations of Brainerd-Robinson coefficients were made for all identified ceramic types and for just the black-on-white decorated types (Table 1).

TABLE 1 Brainerd-Robinson Coefficients of Ceramic Assemblage Similarity (from Rautman 1993).

Ware Categories	Ceramic Samples		
	Kite Site - Sierra Blanca	Kite Site - Rio Abajo	Sierra Blanca - Rio Abajo
All wares	149.3	96.3	100.7
Painted wares	61.6	28.1	20.4
Black-on-white	168.9	30.1	23.6

Ware categories used in the first analysis include brownwares, whitewares (including both decorated and undecorated sherds from black-on-white vesse;s), graywares, redwares, polychromes, glazed sherds, and terracotta sherds. Unidentified ware types were omitted from the analysis. Data from the Kite Site were compared with these from the Sierra Blancas and the Rio Abajo region. The relationship between the Sierra Blanca and Rio Abajo region assemblages was also investigated for comparison (Table 1). Since no particular interaction between these last two areas is proposed, this measurement can represent a baseline value, providing some idea of what the coefficient would be if one compared any randomly selected pair of regions.

As expected, there is greater similarity in ceramic assemblages in the Kite Site and Sierra Blanca regions, with the value of the Brainerd-Robinson coefficient (BR=149.3) nearly 50 percent greater than for any other dyad (see Table 1). In comparison, ceramic assemblage similarity between the Kite Site and Rio Abajo region (BR=96.3) is nearly the same as for the Rio Abajo-Sierra Blanca dyad (BR=100.7), despite the greater distances between the locations of the sites in the "baseline dyad."

When only painted ceramic types are considered—that is, when brownwares are omitted from the analysis—the Brainerd-Robinson coefficient is not particularly high for any combination of pairs (this is most likely due to the wide variety of black-on-white types at the Kite Site, and the greater representation of polychromes in the Sierra Blancas). Still, the value for the Kite Site-Sierra Blanca combination (61.6) is three time the value of the Brainerd-Robinson coefficient for the other two combinations (BR=28.4 and 20.4).

The most marked difference between the Kite Site-Sierra Blanca dyad and the other possible regional combinations occurs when only black-on-white ceramic types are considered. The value of the coefficient for this dyad is 168.9, which is nearly six times greater than the very low values (30.1 and 23.6) for the other possible pairings.

In this case, the Kite Site and the Sierra Blanca region are quite similar (BR=168.9), despite a greater variety of black-on-white types at the Kite Site. The very high proportional representation of Chupadero Black-on-white in both areas is apparently the single most important factor affecting the similarity of these two ceramic assemblages.

Another factor contributing to ceramic assemblage similarity is the simple temporal overlap in site occupation. This possibility is acknowledged here, and may in fact contribute to the relative high proportion of Socorro Black-on-white pottery at the Kite Site (Socorro Black-on-white pottery is the second most prevalent painted type). Comparison of the manufacturing dates of all painted pottery types with the times of occupation of the sites, however, shows that this factor is not sufficient to account for the observed patterning of the ceramic assemblages in these three areas (Rautman 1990, 1993).

RECIPROCITY AMONG NETWORK PARTICIPANTS

The climatic model presented above considers only the structure of social networks. In this section I consider some of the implications of this structure on the operation of the network: specifically, the operation of reciprocal relationships within such a network.

Social relationships within a regional network are considered to involve a number of dyadic relationships that are created, sustained, and expressed by exchanges of food, objects, and information within a system of balanced reciprocity operating outside the local group (Wiessner 1986).

According to Sahlins (1972), relationships based on balanced exchanges are inherently fragile and unstable over time, at least partly because they tend toward self-liquidation (p. 223). On the one hand, there is always the danger that one partner will renege and break off the trade relationship; on the other, continued balanced dealings between distant parties reduces the social distance between them (p. 223) and increases the chances that the relationship will develop into a more generalized exchange relationship (such as those occurring among close kin) in the future.

Social relationships within a risk-reducing exchange network may thus require certain conditions to maintain network continuity. For example, conditions such as those involved in asymmetrical exchanges create, in Sahlins' terms, a "shadow of indebtedness" that contributes to network continuity. In fact, it is "socially critical" that exchanges never become evenly balanced (Sahlins 1972), since as long as the calculation of social accounts is uneven, then there exists the expectation of further associations, and the hope of further payment.

Partners in exchanges may provide for the continuity of exchange relationships by maintaining asymmetry in the amount of "payment," in the character of the exchanged items, or in the timing of exchanges. Thus, exchanges can be bilateral (involving both parties), but differing in the amounts of like items exchanged, or in the substance of exchanged items—for example, food might be exchanged for material items. Alternatively, the exchanges might be more typically a unilateral "gift," which will be repaid, but unevenly, at some unspecified time in the future.

However, too much asymmetry in an exchange relationship will also contribute to network instability. In such a system, there is a high probability of one partner reneging, since the social sanctions for continuing the relationship are presumably not as stringent as those existing in the context of generalized reciprocity within the kin group.

A social network that reduces the risk of resource stress by use of social storage, or storage of reciprocal obligations, is expected to continue only if there is a reasonable balance or equilibrium maintained between giver and receiver populations. Continued or marked asymmetry of exchange attendant upon asymmetry of the giver-receiver relationship may destabilize the network, as the social costs of maintaining contact with spatially disparate groups outweigh expected benefits. Changes in the degree of asymmetry of exchanges within social networks have been suggested to be important in the development of social ranking (Halstead and O'Shea 1982) and also in the development of large-scale regional political alliances (Plog 1984).

The operation of reciprocal exchanges within a social network therefore involves some social strategies that operate to maintain a certain degree of exchange asymmetry in order to ensure network continuity, yet other social and natural factors (such as demographic change or random climatic fluctuations) may create excessive and disruptive asymmetry of interaction.

In this example I test whether the social interactions postulated between the Gran Quivira area and the Ancho/Corona area could be expected to be relatively stable though time. I evaluate here how human groups based in the Ancho/Corona area might view participation in such a relationship.

Ideally, of course, investigation of climatic patterning in central New Mexico might include evaluation of all weather stations in comparison with all other weather stations. Such an analysis, however, is beyond the scope of this study. To evaluate possible sources of asymmetry in the relationship between the two areas at issue—the Salinas area and the Sierra Blanca area—it is necessary only to examine the pairs of weather station areas involved.

At Corona, the mean annual precipitation over 64 years of record is 38 cm., and there were 34 years during which precipitation at Corona was less than the station mean. During these "poor years" at Corona, conditions at the Gran Quivira weather station were average or better than average for only six years. Since the available climatic records at these two stations overlap for only 18 years, the percentage of years for which Gran Quivira is experiencing conditions of "no stress" (the station annual precipitation is equal to or greater than the station mean) is about 33 percent.

This value for the Corona-to-Gran Quivira relationship is considerably higher than that for the Ancho-to-Gran Quivira comparison. During stress years at Ancho ($n = 26$, out of 53 years of weather records), conditions at Gran Quivira are at least average only 18 percent of the time.

The implications of these data for the structure of social networks and coping strategies in central New Mexico obviously depends on a number of factors, among them the value of using these data as probability estimates for avoiding resource stress. It we assume, however, that in general preferred alternative resource areas will be those that consistently experience conditions of "no stress," then we can assess how human groups in these other areas might evaluate the potential utility of participation in social networks that extend to the Gran Quivira area.

For human groups based at the Gran Quivira area, for example, Corona appears to be an appealing alternative resource area during 30 percent of poor years; human groups around Corona would apparently be nearly equally interested in maintaining access to the Gran Quivira area (among others), since when the Corona area experiences stress, the Gran Quivira area is experiencing at least average conditions nearly 33 percent of the time. These data show that conditions at these two stations are in fact reasonably complementary.

The same relationship of roughly equal complementarity holds true for the Gran Quivira/Ancho pair, although prehistoric peoples located near Ancho may be slightly less inclined to maintain long-term reciprocal relationships with groups in the area around Gran Quivira. This analysis of course cannot tell us the threshold percentage level below which reciprocal relationships are not expected to endure, but the data show little evidence of a dramatic disparity in complementarity between the selected station pairs. I suggest that a lack of complementarity in patterning of the climatic variables considered here is therefore unlikely to be a source of asymmetry in the relationship between network participants in these areas.

The same data highlight another point: groups in both the Kite Site area and the Sierra Blancas would do well to participate in exchanges with a variety of other groups in the region. The values cited above represent the highest observed values in central New Mexico; nevertheless, there are obviously numerous years during which social contacts in these proposed areas would prove to be inadequate to mitigate possible local resource stress. Interaction and exchange with groups in the specified alternative resource areas are therefore expected to be a particularly important strategy for coping with local resource stress, but by no means the only one.

DISCUSSION AND CONCLUSIONS

For the prehistoric inhabitants at the Kite Site, social interactions with hunter-gatherers or part-agricultural groups near Corona and Ancho could have provided access to resources in order to reduced the risk of one kind of resource stress: that which may have occurred during years of below-average productivity. It is not necessary to assume that every year of below-average rainfall would have been particularly "stressful" in any real sense for populations at the Kite Site, nor is it necessary to assume that social relations with groups near Corona and Ancho would suffice to alleviate all such years that did prove to be "stressful." The model simply predicts that the spatial patterning and extent of relatively frequently occurring, locally poor conditions apparently are the dominant variables defining the nature of the risk that observed social interactions were structured, in part, to buffer.

This predicted importance of social relationships near Corona and Ancho was a surprising result, and not at all intuitively obvious. At the beginning of these analyses, it seemed likely that differences in elevation would play a large role in structuring human use of the region, simply because there is such a strong correlation between elevation and climatic factors such as precipitation and frost-free season length. It is interesting to note that differences in elevation alone, or even general measures of productivity such as precipitation levels and variation, could not by themselves adequately predict the apparent importance of the proposed alternative resource areas during the Pithouse Period.

The fact of social interactions and possible ceramic exchange (or exchange of ideas regarding ceramic decoration) between groups in the Gran Quivira area and those near the Sierra Blancas is itself not particularly surprising, since the Chupadero Black-on-white ceramic style is considered to be "local" to both areas (e.g., Mera 1931). The model, however, has implications beyond its ability to predict patterns in the archaeological record. After all, in this case, a general assumption of interaction had already been made, based simply on the occurrence of the same "local" pottery in both areas (Mera 1931). The model has its greatest value in providing criteria for explaining observed patterning in the archaeological record. It provides an independent measure—a null hypothesis—that enables us to evaluate where and under what conditions different types of environmental variability may have posed problems for prehistoric cultural groups in the region, and to evaluate the potential success of cultural responses, such as the development of social networks, in coping with risks attendant upon that level of variability.

The type of environmental variability investigated here—interannual variation around the mean—is by no means expected to prove especially stressful to inhabitants at the Kite Site, particularly when one considers how frequently "poor" years occur. For example, it is doubtful that the social network utilized to "cope" with this level of lowered productivity would be treated as a special "coping strategy" that must be implemented in times of crisis. Rather, strategies to maintain access to resources of different regions are expected to constitute the ordinary pattern of life for the Kite Site inhabitants. These strategies, however, may prove to be particularly useful, or perhaps hopelessly inadequate, for buffering other types or levels of stress. The archaeological evidence of such coping strategies can, however, indicate the types of risk the observed sociocultural system was structured to meet.

Risks associated with environmental variability are of course not defined only by climatic factors, but are also constituted by the social, economic, and technological variables that may affect the impact of a given type of environmental variability on a particular social group, the range of strategies that would be potentially available for coping with it, and the social disruption involved in employing a given coping strategy in a specified cultural context. For example, some situations simply occur too infrequently to justify social preparation for them. Other environmental hazards may be devastating to farmers, for example, but may not particularly affect hunter-gatherers in the same area. One can imagine how changes in productive organization at the Kite Site—for example, increasing reliance on maize agriculture—could cause the inhabitants of the Kite Site to reevaluate the nature of the risks that they would be encountering, and the utility of extant social relationships in coping with those risks.

Changes in other areas—for example, a change in technological organization in the Rio Abajo region to include construction of water-retention facilities such as gravel-mulched fields (described by Anschutz [1994] in the northern Rio Grande)—might also cause the inhabitants of the Kite Site to re-evaluate the relative utility of that region in buffering stress caused by low interannual productivity in the Gran Quivira region. This example points out particularly the role of social factors involved in maintenance of interaction, factors that affect the social costs to the participants in comparison to the benefits provided.

Archaeologists need improved chronologies and paleoclimatic records so that correlations between climatic patterns and sociocultural changes can be identified more accurately and at a smaller temporal scale. But we also need better models of the ways in which climate and human actions are interrelated. This study addresses particularly how climatic variability affected prehistoric hunter-gatherers at one site in central New Mexico. More generally, however, this kind of modeling helps identify the climatic and sociocultural variables that affect the organization, operation, and temporal stability of social networks among both living and prehistoric societies in other areas as well.

REFERENCES

Anschuetz, K. F.
1994 Floods, Frosts, Hot Rocks, and Systems of Water Management: Reconsidering Classic Period Anasazi Agricultural Practices in the Northern Rio Grande. Paper presented at the 59th Annual Meeting of the Society for American Archaeology, Anaheim, CA.

Bailey, R. C., and N. R. Peacock
1988 Efe Pygmies of Northeast Zaire: Subsistence Strategies in the Ituri Forest. In *Coping with Uncertainty in Food Supply*, edited by I. de Garine and G. A. Harrison, pp. 88–117. Clarendon Press, Oxford.

Bean, L. J.
1972 *Mukat's People: The Cahuilla Indians of Southern California.* University of California Press, Berkeley.

Brainerd, G. W.
1951 The Place of Chronological Ordering in Archaeological Analysis. *American Antiquity* 16:301–313.

Braun, D. P., and S. Plog
1982 Evolution of "Tribal" Social Networks: Theory and Prehistoric North American Evidence. *American Antiquity* 47:504–525.

Cashdan, E. A. (editor)
1990 *Risk and Uncertainty in Tribal and Peasant Economies.* Westview Press, Boulder.

Clarke, W. C.
1985 Scales of Climatic Impacts. *Climatic Change* 7:5–28.

Colson, E.
1979 In Good Years and Bad: Food Strategies in Self-Reliant Societies. *Journal of Anthropological Research* 35:18–29.
Cordell, L. S., and G. J. Gumerman (editors)
1989 *Dynamics of Southwest Prehistory*. Smithsonian Institution Press, Washington, DC.
Cordell, L. S., and F. Plog
1979 Escaping the Confines of Normative Thought: A Re-evaluation of Puebloan Prehistory. *American Antiquity* 44:405–429.
Cronen, W.
1983 *Changes in the Land: Indians, Colonists and the Ecology of New England.* Hill and Wang, New York.
Dean, J. S.
1995 Demography, Environment, and Subsistence Stress. This volume.
Dean, J. S. and W. J. Robinson
1977 *Dendroclimatic Variability in the American Southwest, A.D. 680 to 1970.* National Technical Information Service, Springfield, VA.
1978 *Expanded Tree-Ring Chronologies for the Southwestern United States.* Laboratory of Tree-Ring Research, Chronology Series III. University of Arizona, Tucson.
Dean, J. S., R. C. Euler, G. J. Gumerman, F. Plog, R. H. Hevly, and T. N. V. Karlstrom
1985 Human Behavior, Demography, and Paleoenvironment on the Colorado Plateaus. *American Antiquity* 50:537–554.
Ford, R. I.
1972 Barter, Gift, or Violence: an Analysis of Tewa Intertribal Exchange. In *Social Exchange and Interaction*, edited by E. Wilmsen, pp. 21–46. Anthropological Papers 46. Museum of Anthropology, University of Michigan, Ann Arbor.
Halstead, P., and J. M. O'Shea
1982 A Friend in Need is a Friend Indeed: Social Storage and the Origins of Social Ranking. In *Ranking, Resource, and Exchange*, edited by C. Renfrew and S. Shennan. Cambridge University Press, Cambridge.
Halstead, P., and J. M. O'Shea (editors)
1989 *Bad Year Economics: Cultural Responses to Risk and Uncertainty.* Cambridge University Press, Cambridge.
Hantman, J. L., and S. Plog
1982 The Relationship of Stylistic Similarity to Patterns of Material Exchange. In *Contexts for Prehistoric Exchange*, edited by J. E. Ericson and T. K. Earle, pp. 237–264. Academic Press, New York.
Hegmon, M.
1995 Variability in Food Production, Strategies of Storage and Sharing, and the Pithouse-to-Pueblo Transition in the Northern Southwest. This volume.

Kelley, J. H.
1984 The Archaeology of the Sierra Blanca Region of Southeastern New Mexico. Anthropological Papers 74. Museum of Anthropology, University of Michigan, Ann Arbor.

Marshall, M. P., and H. J. Walt
1984 *Rio Abajo: Prehistory and History of a Rio Grande Province.* New Mexico Historic Preservation Bureau, Santa Fe.

Mera, H. P.
1931 *Chupadero Black on White.* New Mexico Archaeological Survey, Technical Series, Bulletin 1. Laboratory of Anthropology, Santa Fe.

Minc, L. D.
1986 Scarcity and Survival: The Role of Oral Tradition in Mediating Subsistence Crises. *Journal of Anthropological Archaeology* 5:39–113.

Minnis, P. E.
1985 *Social Adaptation to Food Stress: A Prehistoric Southwestern Example.* University of Chicago Press, Chicago.
1995 Notes on Economic Uncertainty and Human Behavior in the Prehistoric North American Southwest. This volume.

Nelson, M.
1995 Technological Strategies Responsive to Subsistence Stress. This volume.

Ortiz, S.
1990 Uncertainty Reducing Strategies and Unsteady States: Labor Contracts in Coffee Agriculture. In *Risk and Uncertainty in Tribal and Peasant Economies*, edited by E. Cashdan, pp. 303–319. Westview Press, Boulder.

O'Shea, J. M.
1981 Coping with Scarcity: Exchange and Social Storage. In *Economic Archaeology*, edited by A. Sheridan and G. Bailey, pp. 167–183. International Series 96. British Archaeological Reports.

O'Shea, J. M. and P. Halstead
1989 Conclusions. In *Bad Year Economics*, edited by P. Halstead and J. M. O'Shea, pp 123–126. Cambridge University Press, Cambridge.

Plog, F.
1983 Political and Economic Alliance on the Colorado Plateaus, A.D. 600–1450. In *Advances in World Archaeology* vol. 2, edited by F. Wendorf and A. Close, pp. 289–330. Academic Press, New York.
1984 Exchange, Tribes, and Alliances: The Northern Southwest. *American Archeology* 4:217–223.

Plog, S.
1976 Measurement of Prehistoric Interaction Between Communities. In *The Early Mesoamerican Village*, edited by K. V. Flannery, pp. 255–272. Academic Press, New York.

Plog, S., and J. L. Hantman
1990 Chronology Construction and the Study of Prehistoric Culture Change. *Journal of Field Archaeology* 17:439–456.

Plog, S., and S. Powell
1984 Patterns of Culture Change: Alternative Interpretations. In *Papers on the Archaeology of Black Mesa, Arizona, Volume II*, edited by S. Plog and S. Powell, pp. 209–216. Southern Illinois University Press, Carbondale.

Rautman, A. E.
1990 *The Environmental Context of Decision-Making: Coping Strategies Among Prehistoric Cultivators in Central New Mexico.* Ph.D. dissertation, University of Michigan. University Microfilms, Ann Arbor.
1993 Resource Variability, Risk, and the Structure of Social Networks: An Example from the Prehistoric Southwest. *American Antiquity* 58:403–424.

Renfrew, C.
1975 Trade as Action at a Distance: Questions of Integration and Communication. In *Ancient Civilizations and Trade*, edited by J. Sabloff and C. C. Lamberg-Karlovsky, pp. 4–39. University of New Mexico Press, Albuquerque.

Robinson, W. S.
1951 A Method for Chronologically Ordering Archaeological Deposits. *American Antiquity* 16:293–301.

Sahlins, M.
1972 *Stone Age Economics.* Aldine, Chicago.

Scudder, T.
1962 *The Ecology of the Gwembe Tonga.* Rhodes-Livingstone Institute, Manchester.

U. S. Department of Commerce
1917–1980 *Climatological Data: Annual Summaries, New Mexico.* National Oceanic and Atmospheric Administration Environmental Data Service, Washington, D.C.

Upham, S.
1982 *Polities and Power: An Economic and Political History of the Western Pueblo.* Academic Press, New York.

Wiens, J. A., J. F. Addicott, T. J. Case, and J. Diamond
1986 Overview: The Importance of Spatial and Temporal Scale in Ecological Investigations. In *Community Ecology*, edited by J. Diamond and T. J. Case, Chapter 8. Harper and Row, New York.

Wiessner, P.
1977 *Hxaro: A Regional System of Reciprocity for Reducing Risk Among the !Kung San.* Ph.D. dissertation, University of Michigan. University Microfilms, Ann Arbor.
1986 !Kung San Networks in a Generational Perspective. In *The Past and Future of !Kung Ethnography: Critical Reflections and Symbolic Perspectives*, edited by M. Biesele with R. Gordon and R. Lee, pp. 103–135. Helmut Buske Verlag, Hamburg.

Winterhalder, B.
1980 Environmental Analysis in Human Evolution and Adaptation Research. *Human Ecology* 8:135–170.

Wobst, H. M.
1974 Boundary Conditions for Paleolithic Social Systems: A Simulation Approach. *American Antiquity* 39:147–178.

Yengoyan, A. A.
1972 Ritual and Exchange in Aboriginal Australia: an Adaptive Interpretation of Male Initiation Rites. In *Social Exchange and Interaction*, edited by E. N. Wilmsen, pp. 5–10. Anthropological Papers 46. Museum of Anthropology, University of Michigan, Ann Arbor.

Michelle Hegmon
Department of Sociology and Anthropology, New Mexico State University, Las Cruces, NM 88003

Variability in Food Production, Strategies of Storage and Sharing, and the Pithouse-to-Pueblo Transition in the Northern Southwest

In this chapter I attempt to understand human social responses to resource stress and economic uncertainty among food producers in the Southwest. Specifically, my focus is on the effects of different strategies of sharing or not sharing food among horticulturalists in an environment with a high level of high-frequency variability. The results are applied to interpret changes in architecture associated with what is generally called the pithouse-to-pueblo transition, focusing on changes from communal to private storage.

THEORETICAL BACKGROUND

One of the many stimulating aspects of the Santa Fe Institute workshop was the opportunity to explore and compare a number of different theoretical perspectives, ranging from selectionist to postprocessual. This paper takes a somewhat eclectic approach, drawing from various perspectives as seems appropriate. Although the form of explanation used here is probably best classed as processual, I also rely heavily on concepts developed in Marxist approaches to anthropology (see McGuire [1993] and Trigger [1993] for recent reviews).

Specifically, in considering the relationship between food production and social organization, I am concerned with the *mode of production.* A mode of production includes both the means for making a living (mode of subsistence) and the relationships involved: "Who owns these means, how is production organized, who controls the product and how is it distributed, and who consumes what part of it?" (Leacock and Lee 1982:7; see Ingold [1988:273–276] for further discussion of modes of production in anthropological analysis; see also Harris [1983] and Himmelweit [1983] for stricter Marxist definitions). The concept of a mode of production, because it is often associated with the structural Marxism of the 1970s (e.g., Meillassoux 1972, 1973; Sahlins 1972), is not particularly popular among Marxist archaeologists today (see McGuire 1993:110). Still, for my purposes, it is useful for archaeological interpretation because it provides a means of conceptualizing social transformations in economic terms, even if those transformations do not involve obvious changes in production technology or subsistence practices.

The perspective provided by considering the mode of production is particularly applicable to understanding certain changes in the prehistoric Southwest. That is, recent evidence suggests that the degree of dependence on food production and/or corn changed relatively little during periods of profound social change. At least three separate analyses of material including coprolites, settlement patterns, bone chemistry, and midden remains (Decker and Tieszen 1989; Matson and Chisholm 1991; Minnis 1989) conclude that Anasazi reliance on corn changed relatively little from Basketmaker II or III through Pueblo III times (i.e., from about A.D. 400 to 1300). Clearly a great deal changed through these centuries, including the development of complex water-control and irrigation techniques used for corn production, as well as social elaboration, aggregation, sedentism, architecture, etc. The apparent stability of corn consumption suggests that these changes are best understood not merely in terms of subsistence strategies, but rather in terms of the social relations that are part of the mode of production.

The focus here is on the storage and distribution of crops and the social relations involved in those practices. Given a certain method of producing food in a variable environment (i.e., given an unchanging mode of subsistence), what are the effects of different social organizational forms (i.e., different modes of production)? A number of authors have developed perspectives and methods of classification useful for understanding the social relations of storage, sharing, and exchange.

Sahlins (1972:193–194) draws a distinction between generalized reciprocity (which Lee [1988:258] likens to primitive communism) and balanced reciprocity. In generalized reciprocity a gift is given with few expectations regarding when and even if it will be returned. Because the social outweighs the material, there are few clear-cut obligations involved in generalized reciprocity. Balanced reciprocity, in contrast, does involve clear-cut obligations for return, and often takes the form of direct exchange. In these cases, social relations are dependent on the transfer of goods. Sahlins (1972:217–218) suggests that, very generally, food is most often given in the context of generalized reciprocity and other goods in balanced reciprocity. He also notes that tribal societies tend to have more balanced reciprocity than bands, possibly because tribes often have more nonfood exchange and because in tribal societies there is more exchange (of food and other goods) with socially distant persons.

Meillassoux (1972, 1973) examines the mode of exploitation of the land among hunter-gatherers and agriculturalists, comparing the Mbuti and the Bantu. He suggests that while in hunting (and presumably gathering) "the act of circulation, as of production, is instantaneous" (1973:194); agricultural production necessitates prolonged and continuous cooperation and thus results in binding ties, including inter-generational obligations. Thus agriculture involves a mode of production in which some people maintain control over others. Agriculturalists (in contrast to hunter-gatherers) tend to have kin-based rather than geographically defined groups, more long-term binding ties, and more concern with the past and future, including an emphasis on ancestors and descent groups as well as control of women and reproduction.

Woodburn (1982; Barnard and Woodburn 1988) contrasts immediate and delayed-return strategies. Some foragers get an almost immediate return on their labor, while returns are more delayed for food producers, as well as other foragers and more complex hunter-gatherers. Woodburn argues that immediate return is associated with strongly egalitarian societies, while delayed return often involves more competition and social differences.

Ingold (1983, 1986:198–221) argues that the critical factor for classifying social forms is "whether or not people are bound to one another by enduring relations in respect to the control and distribution of means of subsistence" (1983:553). A critical factor in this distinction is what he calls social storage, which involves the appropriation of resources by certain individuals and thus inegalitarian relationships.[1] Ingold argues that social storage, and the inequality it engenders, are associated with food-producing as well as some hunter-gatherer economies, while other hunter-gatherers (primarily those that would be called foragers with immediate return) lack social storage. In societies of the latter type, sharing is determined primarily by ecological considerations and the need of all to have access to collective resources. That is, not everything is always shared (meat from large game is generally shared much more extensively than plant foods), but an absence of sharing does not imply "curtailment of sharing as a *social principle of collective appropriation*" (1983:562, emphasis in original). In contrast, social storage "does represent the direct negation of sharing," and often involves private or individual storage of goods (1983:563).

Wills (1991, 1992) contrasts communal and household systems. A communal system is characterized by generalized reciprocity and economic strategies that are

[1]Ingold's use of the term social storage is very different from O'Shea's (1981). By social storage, O'Shea (1981:169) refers to the exchange of "food for some non-food token with at least the implicit understanding that such tokens can later be re-exchanged for food."

risk-averse and have diminishing returns (i.e., an increase in investment does not lead to a proportional increase in returns). In contrast, household systems involve less generalized reciprocity and follow risk-prone, increasing-returns strategies. Thus in a communal system, surplus production is inhibited, both by the diminishing returns economy and by the demands of generalized reciprocity. In contrast, household systems can generate surplus, which is often kept by individual households. Wills applies his model to early food producers in the Southwest, comparing pithouse occupations at the Mogollon/SU Site (near Reserve, New Mexico) and Shabik'eschee Village, a Basketmaker III site in Chaco Canyon. He suggests that the SU Site, which has large pithouses with numerous internal (and therefore private) storage pits, represents a household system, while Shabik'eschee, which has external storage cists, more likely represents a communal system.

The list of comparisons between different modes of production could go on (e.g., Bender 1990; Kelly 1991; Price and Brown 1985; Testart 1982). All the classification schemes are slightly different, though one theme that is common to most is that food-producing societies tend to have less egalitarian social relations and less even distribution of resources than foragers. It is this relationship that is considered here. Specifically, I explore whether some component of food production and the pattern of yields might explain the relatively inegalitarian relations observed in food-producing societies.

Although the anthropological literature on sharing and exchange is extensive and growing fast, there seems to be little consensus regarding the circumstances under which sharing will increase or decrease. A number of scholars (e.g., Braun and Plog 1982; Hames 1990; Heinen and Ruddle 1974; Kaplan and Hill 1985; Sahlins 1972; Winterhalder 1986a) consider sharing to be useful as a means of reducing risk (usually measured as variance or unpredictability). That is, in Evans-Pritchard's (1940:85) terms, "it is scarcity, not sufficiency, that makes people generous." However, recent work suggests that sharing, as only one of a number of strategies for reducing risk, is not always advantageous (e.g., Cashdan 1985; Kelly 1991; Smith 1988; Winterhalder 1986a, 1990). Some (e.g., Cadelina 1982; Kohler and Van West, this volume; Van West and Kohler 1992) argue that especially among food producers, sharing is expected to be limited in times of scarcity.

The research presented here attempts to understand if, when, and how sharing would be advantageous to food producers in the environment of the northern Southwest. Given the topic of this conference and the nature of the area, the focus is on production in a highly variable environment, that is, an environment with a high degree of microtopographic variation and a large degree of year-to-year variability.

The remainder of the paper is in two parts. The first is based on a computer simulation of corn production using data derived from studies of the twentieth-century Hopi. The details of this simulation and issues of risk and variation are examined in Hegmon (1989), and a different version of the simulation is presented in Hegmon (1991). The purpose of the present work is to combine some of these findings and extend them (in the latter part) to understanding architectural and social transitions in the prehistoric Southwest.

THE SIMULATION MODEL

THE HOPI

The Hopi are pueblo horticulturalists/agriculturalists living in the mesa country of northeastern Arizona. The ethnographic present for this study is the early part of this century, and most of the data for this study are from accounts of Second Mesa (Beaglehole 1937; Forde 1931; also Eggan [1950] and Titiev [1944] for First and Third Mesas; and Bradfield [1971]; Connelly [1979]; Hack [1942]; Kennard [1979]; and Whiteley [1985a, 1985b] for the Hopi in general).

Very briefly, for the purposes of this study, Hopi social organization can be understood in terms of the obligations that are involved in various social relationships, following Connelly (1979). The household is the basic unit of everyday activities, including production and consumption (Beaglehole 1937:5). Descent groups (matrilineages and clans) are to some extent complementary to households since they are basic units in ritual and sometimes inheritance. Classic interpretations of the Hopi (e.g., Eggan 1950) consider descent groups, particularly clans, to be corporate land-owning groups, though this interpretation has been strongly contested by Whiteley (1985a, 1985b). Whiteley argues that descent groups are not necessarily unambiguous permanent units with absolute claims to land. For the purpose of this research, however, consideration of these larger-than-household groups is still important because of their implications for food distribution and sharing, regardless of their permanence or corporate nature. Connelly (1979) makes clear that within households, matrilineages, and clans there are strong obligations to share or exchange food, labor, and ritual services. Above the level of the clan, obligatory relationships are much weaker.

Hopi subsistence is broadly based, but corn is *the* staple. Corn is important not just as food, but also in social and ritual contexts. Furthermore, because other foods can substitute for corn in subsistence but not in socio-religious contexts, subsistence failure will be felt socially before starvation sets in. Thus the social emphasis on corn may trigger early or timely responses to subsistence stress (see Minnis, this volume). Because corn is so closely tied to social relationships, this study focuses only on corn.

The conditions on the Hopi Mesa are just sufficient for growing corn, though rainfall and growing season length are unpredictable and often limit productivity (Hack 1942; though *contra* Adams [1979]). The Hopi adapt to this difficult environment in two general ways: (1) with techniques developed to take advantage of microtopographic variation; and (2) with a system that mitigates the effects of failure.

Hopi corn is well suited to arid conditions, and the Hopi grow several varieties, each with slightly different advantages and disadvantages (Brown et al. 1952; Collins 1914). The Hopi also plant different kinds of fields in different microtopographic settings (Forde 1931; Hack 1942). Most common are *akchin* fields, which are set at the mouth of an arroyo and watered by runoff (see Hack 1942:28). Other kinds of irrigated fields include those planted on the terraces of large arroyos, in the bottom of small arroyos, and on dams built in small arroyos (*trinchera* fields) (see Hack 1942:30). In addition, the Hopi situate some fields on dunes to take advantage of seepage through the massive sandstone reservoir of Black Mesa (Hack 1942:32–34). These different fields are advantageous in different conditions. For example, arroyo bottom fields shelter plants from cold and provide plenty of moisture, but are susceptible to flooding. Dune seepage fields do not receive large quantities of moisture, but are rarely flooded.

Hopi flexibility increases the chance that although not all fields yield each year, what is produced will be widely available (Plog 1978). The different kinds of fields can take advantage of various levels of moisture. Planting dates are spread over three to four months, so regardless of the timing of frosts, some fields will yield (Bradfield 1971; Titiev 1938). Fields are scattered and most households have several fields. Furthermore, corn is also shared among households and is stored from one year to the next. Thus there is a good chance that each household will have access to some corn each year. Unfortunately, although ethnographic accounts clearly document sharing (e.g., Kennard 1979:561), they contain little information regarding the organization and extent of the sharing.

THE SIMULATION OF CORN PRODUCTION

Although ethnographic data on sharing strategies are not available, computer simulation can be used to consider the effects of various strategies of sharing. Data on Hopi subsistence practices are used to develop a simulation model to investigate the relationship between corn production and distribution.food production, corn The simulation is not intended to model everything the Hopi do. Instead, it simulates highly variable production based on Hopi data and thus facilitates the investigation of the relationships among certain variables. The simulation is constructed in two parts. The first models corn production and consumption at the household level. The second considers the consequences of different strategies of interhousehold sharing.

For purposes of the simulation, all households have the same composition, production capability, and nutritional needs. Each comprises an elderly couple, a middle-aged couple, 2 seven- to ten-year-old children, and a four- to six-year-old child, and needs 5,584,500 kcal.[2] annually. Corn provides 71 percent of the caloric needs, so each household needs 1017 kg. of corn annually.[3]

Each household farms 3.15 ha. of corn, divided among three fields. Two of these fields are akchin, the third is a randomly selected different type (i.e., terrace,

[2] Based on recommended daily dietary allowances of the Food and Nutrition Board, National Academy of Sciences, National Research Council.

[3] In her simulation involving Rio Grande Pueblos, Spielmann (1982) estimated that 75 percent of caloric needs were met by corn; Wetterstrom (1976) used estimates of 71.1 to 71.5 in her study of Arroyo Hondo Pueblo. Corn provides approximately 3.9 kcal. per gram (Wetterstrom 1976:255).

trinchera, arroyo bottom, or dune seepage). The simulation begins by randomly assigning each household a set of three fields. If the third field is irrigation-based, this field comprises only 10 percent (0.32 ha.) of the household's land, since land around arroyos is limited. If the third field utilizes dune seepage, it comprises 20 percent (0.63 ha.) of the land. No inequalities are built into the simulation, but the different fields are advantageous under different conditions.

For each year the simulation determines the yield of each field and the total yield for each household. The calculation is based upon yields expected under average growing season length and summer rainfall conditions (500 kg./ha. for *akchin* fields, 425 kg./ha. for dune seepage fields, and 667 kg./ha. for other irrigated fields). Deviations from these expected yields are then computed based on three sets of factors: (1) variation in the annual weather conditions; (2) microtopographic variation in the conditions experienced by each field; and (3) damage to the fields by severe storms and insects. The first set of deviations is based on real weather data, the second and third are modeled with computer-generated probabilities.

The effect of annual weather is calculated with respect to the July-August rainfall and the length of the growing season, using data from the Keams Canyon and Jeddito weather stations (Sellers and Hill 1974; U.S. Department of Commerce 1932–1972). These weather data are considered not as raw figures but as percentages of the mean at each station. This practice emphasizes the effect of variation and reduces the effect of different microtopographic conditions at the weather stations and fields.

Variation in the conditions experienced by each field are then modeled in terms of probabilities that a given field experiences the annual weather conditions or some deviation caused by microtopographic variation. For example, an *akchin* field has a 70 percent probability of receiving the annual input of available water, a 5 percent probability of receiving the annual input plus 30 percent, a 5 percent probability of receiving the annual input minus 30 percent, a 10 percent probability of receiving the annual input plus 20 percent, and 10 percent probability of receiving the annual input minus 20 percent. Terrace fields, watered by large arroyos, have a greater chance of receiving more water.

Given the conditions experienced by each field, the simulation calculates the baseline yield for the field, using figures developed by Wetterstrom (1976: Tables 40&66). For example, if the water available to a field is 111 or more percent of the average, the corn yield is 111 percent of the average; if the growing season length is 50 to 75 percent of the average the yield is 67 percent of the average. Under these conditions, a 1-ha. *akchin* field would yield 370 kg. of corn.

Once these baseline yields are calculated, the simulation models damage to the yield due to floods, hailstorms, and grasshoppers. The different field types are differentially susceptible to flooding; all are equally susceptible to hail and grasshoppers. If a field is damaged by one of these factors, its baseline yield is reduced by a certain percentage; in some cases the yield is reduced to nothing.

TABLE 1 Sample of yield output generated by simulation, showing corn produced by Households 1–5 in years 1932–1937. These data are used in the simulation of the different sharing strategies (Tables 2–4).

	year	HH1	HH2	HH3	HH4	HH5
	1932	1226	1286	1714	1140	1262
RAW	1933	768	920	1099	1004	981
YIELDS	1934	788	709	483	821	394
KG.	1935	965	906	709	471	965
CORN	1936	512	608	118	699	1139
	1937	1588	568	1278	1556	710

The simulation of production was run for 2 20-year periods, using weather data from 1932–1951 and 1952–1972 (excluding 1955). One hundred households were simulated during each period. The output consists of the annual summed yields of each household's fields. A sample of five households is shown in Table 1.

SHARING AND DISTRIBUTION STRATEGIES

The data on the annual yield per household are used to examine different strategies of sharing and distribution, specifically strategies of household independence, pooling, and restricted sharing. All three strategies involve the same consumption requirements and storage, but they differ in terms of interhousehold sharing. Independence involves no sharing; restricted sharing involves sharing only household surplus; and pooling involves complete sharing.

Consumption is modeled in terms of households' annual caloric needs, which include 1017 kg. of corn. If a household does not get the requisite 1017 kg., it does not necessarily starve, but it will be unable to meet its ritual and social needs and therefore can be said to fail socially. One poor year can be tolerated, but a household that repeatedly has poor years will be a strain on the social system and will face nutritional problems. Therefore, for these runs of the simulation, a household can survive up to two years in a row in which it cannot meet its needs. However, if a household does not meet its needs three years in a row, it fails to survive in the simulation.

Storage, whether on a household or group level, is part of all three distribution strategies. Perfect storage, with no loss or waste, is assumed in the simulation. Based on test runs with average weather conditions, each household begins the simulation with 266 kg. of corn in storage.

Independent households survive or fail on their own, depending only on the yields of their three fields and their private storage. Each year the yield of a household's fields is added to whatever it has in storage and the total is compared to the needed 1017 kg. If the total is greater than 1017, the household consumes 1017 kg. and the surplus is put into storage. If the total is less than 1017 the household consumes all it has, begins the next year with nothing in storage, and the simulation assigns it one demerit. If the household succeeds in meeting its needs within the next two years, the demerit is dropped. If it has three years in a row in which it fails to meet its needs (and thus accumulates three demerits), it fails to survive in the simulation.

Data from five households are shown in Table 2. Household 1's fields yield 1226 kg. of corn in 1932, so with the 266 kg. it has in storage it has a total of 1492 kg. available. It consumes 1017 and the remainder (475 kg.) goes into storage. Its yield in 1933 is only 768 kg., but with its storage it still has enough (1243 kg.) and puts 226 kg. in storage. In 1933 its yield is again low (788 kg.), and even with the storage it has only 1014 kg. available, less than the needed 1017 kg. It consumes all its corn, receives one demerit, and begins the next year with nothing in storage. In the following two years Household 1 also fails to produce the requisite 1017 kg. It thus accumulates three demerits and fails to survive in the simulation.

The simulation was run with 100 independent households over the 2 20-year periods. Fewer than half (45 and 46 percent) of the households survived over the 20-year periods. This low survival rate is not unexpected, given the emphasis on sharing and interdependence among the Hopi and other food producers.

Pooling involves a high degree of interdependence among households. The households in a group pool all their stores and yields, and their needs are assessed on a group basis. Only groups of five households are considered here; the effect of different group sizes was examined in Hegmon (1991). The simulation of the groups is similar to that of independent households multiplied by five (the same yield data [Table 1] are used).

Data for five households in one group are shown in Table 3. In 1932 the total yield of the five sets of fields in addition to the beginning stores (266 kg./household [hh.]) is 7955 kg. corn, so each household has 1591 available. Each consumes 1017 kg. and the remainder (a total of 2870 kg.) is put in storage. Sufficient corn is available in 1932, 1933, and 1934, but there are shortfalls in 1935 and 1936. Thus the group begins 1937 with two demerits and nothing in storage. However, the total yield for 1937 is 5700, which is enough to meet the 1017 kg./hh. need. Thus, the demerits are eliminated and the group begins 1938 with a small surplus (615 kg. total).

The simulation was run with 100 households (20 groups) over the 2 20-year periods. Pooling increases the survival rate over independence; 11 of the 20 survived the first run and 18 of the 20 survived the second, for an average survival rate of 72 percent. Thus pooling is more successful than independence in the simulation,

TABLE 2 Results of household independence, showing corn available to simulated Households 1–5 in years 1932–1937. Figures are for corn available before consumption, and include annual yield (Table 1) and any corn in storage.

	year	HH1	HH2	HH3	HH4	HH5
CORN	1932	1492	1552	1980	1406	1528
AVAIL.	1933	1243	1455	2062	1393	1492
TO	1934	1014[1]	1147	1528	1197	869[1]
HOUSE-	1935	965[2]	1036	1220	651[1]	965[2]
HOLDS	1936	512[3]	627[1]	321[1]	699[2]	1139
	1937	-0[4]	568[2]	1278	1556	832[1]

[1] Demerit, household failed to meet consumption needs (1017 kg./hh./year).

[2] Two demerits.

[3] Three demerits.

[4] Household accumulated three demerits and was dropped from simulation.

TABLE 3 Results of pooling, showing corn available to simulated Households 1–5 in years 1932–1937. Figures are for corn available after sharing but before consumption. Calculations based on yield data shown in Table 1.

	year	HH1	HH2	HH3	HH4	HH5
CORN	1932	1591	1591	1591	1591	1591
AVAIL.	1933	1529	1529	1529	1529	1529
TO	1934	1151	1151	1151	1151	1151
HOUSE-	1935	938[1]	938[1]	938[1]	938[1]	938[1]
HOLDS	1936	616[2]	616[2]	616[2]	616[2]	616[2]
	1937	1140	1140	1140	1140	1140

[1] Demerit, household failed to meet consumption needs (1017 kg./hh./year).

[2] Two demerits.

TABLE 4 Results of restricted sharing, showing corn available to simulated Households 1–5 in years 1932–1937. Figures are for corn available after sharing but before consumption. Calculations based on yield data shown in Table 1.

	year	HH1	HH2	HH3	HH4	HH5
CORN	1932	1492	1552	1980	1406	1528
AVAIL.	1933	1243	1455	2062	1393	1492
TO	1934	1017	1096	1477	1146	1017
HOUSE-	1935	1006	1017	1017	641	1006
HOLDS	1936	543	639	149	730	1017
	1937	1335	1017	1025	1303	1017

though a survival rate of fewer than three out of four is still somewhat marginal. Detailed examination of the annual data suggests that with pooling, one household that does poorly can pull down an entire group.

Restricted Sharing is intermediate between independence and pooling. With restricted sharing, households share in groups of five, but they do not pool everything. A household first meets its own needs, and only if it has more than enough does it share with other members of the group. Any surplus remaining after sharing goes into the household's private storage. The household, not the group, is the unit of survival. The same yield data as were used for the other strategies are used in the simulation of restricted sharing.

Data for the sample of five households with restricted sharing are shown in Table 4. In 1932 and 1933 all the households have enough corn so there is no sharing (the results are the same as with independence). In 1934 Households 1 and 5 have shortfalls while Households 2, 3, and 4 again have surpluses. Households 2, 3 and 4 each consume the needed 1017 kg. and give Households 1 and 5 enough to meet their needs. The remainder goes into the stores of the households that produced the surplus. In these six years, no household experiences a year in which it does not get the necessary 1017 kg. of corn.

The simulation of households with restricted sharing in groups of five was run with 100 households for both 20-year periods. All survived the first period, and 84 of the 100 survived the second, for an average survival rate of 92 percent. Thus overall it appears that restricted sharing is the most successful of the simulated strategies, though pooling was slightly more advantageous in the second period. Figure 1 compares the strategies graphically.

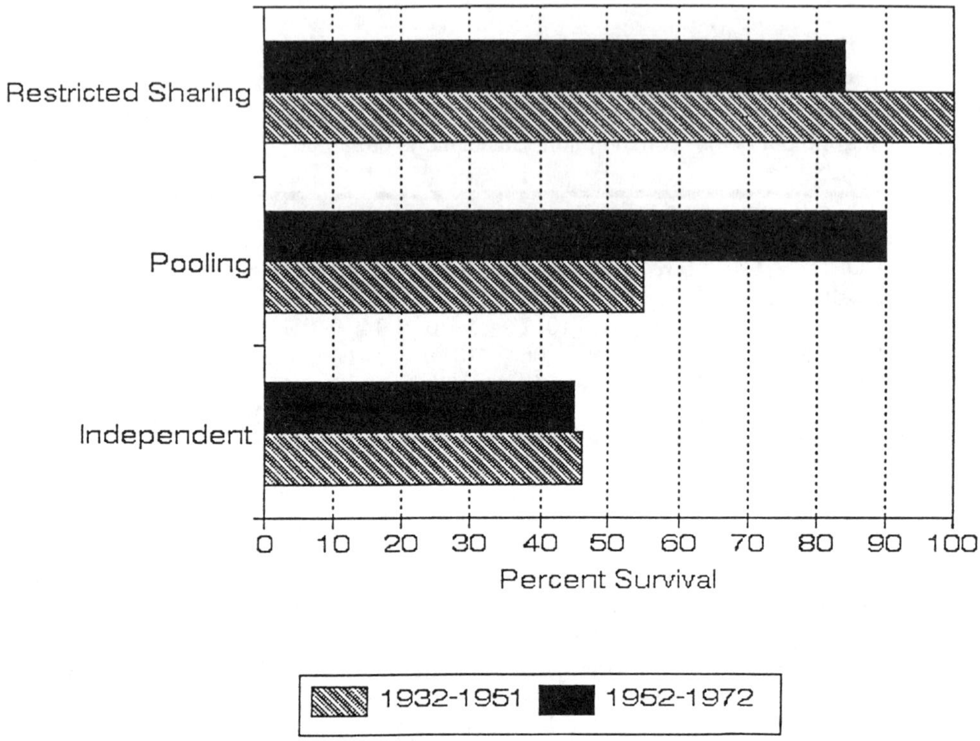

FIGURE 1 Bar chart showing household survival rate for the three sharing strategies in the two time periods.

These results are interesting, though not all that surprising. They suggest that the strategy of restricted sharing, which most closely approximates that used by the Hopi and other small-scale food producers, is the most advantageous. In other words, these data show that what these people have done for a long time makes sense. Given the kinds of yields expected in a highly variable environment like that in the Southwest, some sharing is advantageous, but too much sharing can be a detriment. Below I explore why this might be so, by considering variation in yields and consumption.

VARIATION

The relative success of the three strategies of distribution may be related to their effects on variation in household yields. Some scholars (e.g., Roumasset 1976:16; Wharton 1971:60; Winterhalder 1986a; see also discussion above) have argued that a low degree of variation should decrease risk and thus increase the survival rate. The relationship between variation and failure in the simulation is examined first by considering the mean variation (i.e., the variance), and second by considering the pattern of variation in the different time periods.

The coefficient of variation is used to examine the relationship between the variance and risk (i.e., the chance of failure). Coefficients of variation were calculated for the annual yields of the 100 households and for the amount of corn available to each household each year in the three strategies (after sharing and distribution but before consumption). The mean coefficient of variation for each category for each 20-year period is shown in Table 5.

All three strategies lower the variation inherent in the raw yields, since all involve storage and/or exchange. Of the three strategies, pooling results in the least variation, while independence and restricted sharing have similar and slightly higher coefficients of variation. Clearly, in this case, low variance does not translate into greater success or a higher survival rate. This conclusion parallels recent findings in optimal foraging theory (e.g., Real and Caraco 1986; Smith 1988; Stephens and Charnov 1982; Winterhalder 1986b) as well as work by Kohler and Van West (this volume). A lower variance, achieved by sharing or other means, is not necessarily advantageous; other factors are also relevant. Kohler and Van West argue that cooperation such as food sharing will be most valuable in circumstances of high production coupled with high temporal and spatial variability. Such cooperation will be much less advantageous if productivity is low.

TABLE 5 Average coefficient of variation for the raw yields and for the corn available to each household with the three strategies of distribution for the 2 20–year periods.

Strategy	Period	Coefficient of Variation
Raw	1932–51	.504
Yields	1952–72	.493
Independence	1932–51	.369
	1952–72	.373
Pooling	1932–51	.321
	1952–72	.296
Restricted	1932–51	.380
Sharing	1952–72	.359

Measures of variance (i.e., mean deviation from the mean) are useful, but they clearly do not tell the whole story. To consider the effect of variation further, patterns of variation in yields for the 2 20-year periods were considered (Figure 2). The three strategies were differentially successful in these periods (Figure 1). Independence was nearly equally unsatisfactory in both periods (45 and 46 percent survival rates). Pooling was fairly unsatisfactory during the first period (55 percent survival) but successful (90 percent survival) during the second period. Restricted sharing was satisfactory in both (100 and 84 percent survival), though it was most successful in the first.

The successes and failures can be related to the different patterns of variation present in the two periods. While the average yields and coefficients of variation are roughly similar for the two periods, good and bad years are distributed very differently. In the first period there are four years in a row (1933–36) in which the average yield is less than the requisite 1017 kg., though the average yield is never below 1017 for more than two years in a row in the second period. These results suggest that the key to survival is surviving bad periods; and restricted sharing—with its combination of private storage and restricted sharing—clearly offers the best prospects for weathering such bad periods.

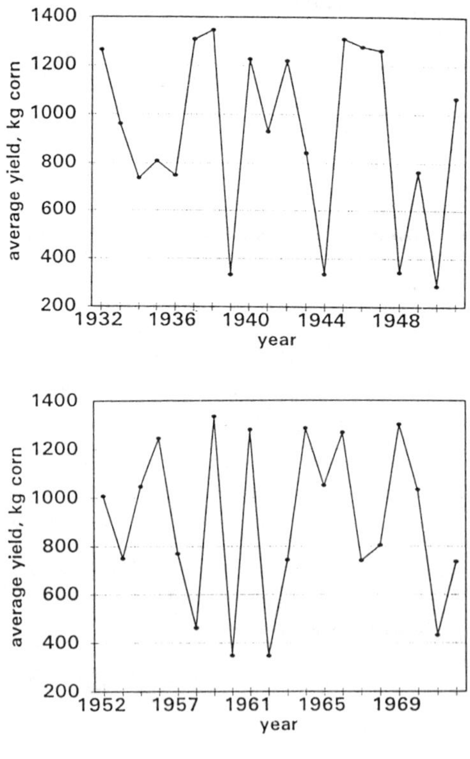

FIGURE 2 Average annual household corn yield for the 2 20–year periods used in the simulation (after Hegmon 1989: Figure 1).

THE PITHOUSE-TO-PUEBLO TRANSITION

These simulation results are relevant to understanding what seems to have been a fundamental change across most of the prehistoric northern Southwest, that is, the pithouse-to-pueblo transition. Very generally, this transition involved a change in residential architecture from pithouses to a combination of pit structures and above-ground pueblos. In much of the northern southwest this change occurred in the eighth and early ninth centuries A.D., that is, between Basketmaker III and Pueblo I in the Pecos Classification, or what Cordell and Gumerman (1989) call the Expansion period. However, in some areas (e.g., parts of the Kayenta region and the Gallina district [Cordell and Plog 1979; Hobler 1974]) pithouses lasted until much later. In many cases, the development of pueblo architecture also appears to be associated with decreases in residential mobility and/or the beginnings of year-round sedentism.

The focus here is on understanding one aspect of the architectural change, specifically, the organization of storage. At many pithouse sites, the primary storage facilities appear to be above-ground cists, granaries, or other structures that are not attached to the pithouses, though pithouse antechambers could also have been used for storage. Such unattached storage facilities are present at large Basketmaker III sites such as Shabik'eschee (Wills 1992; Wills and Windes 1989) and Broken Flute Cave (Hays 1991; Morris 1980), as well as on numerous (and more typical) small sites such as the Tres Bobos Hamlet (5MT4545) near Dolores, Colorado (Brisbin and Varien 1986; see Figure 3) and sites near Allantown, Arizona (Roberts 1939). The placement of these storage facilities in what appear to be communal areas (at least on the large sites) suggests that the storage was relatively communal and might have been associated with a pooling strategy (Wills 1992; Wills and Windes 1989).

In contrast, pueblo sites often comprise double rows of rooms, with smaller storage rooms in the rear and larger habitation rooms in the front (Dean 1969; Hill 1970; Kane 1986; Lightfoot and Etzkorn 1993; see Figure 4). In the Mesa Verde region, early pueblo storage and habitation rooms are almost always attached, while in other areas (such as the Kayenta region) storage and habitation rooms are attached only on the larger sites (e.g., D:11:2030 on Black Mesa [Green et al. 1985]), but are separate on smaller sites (see Hegmon 1994). The attachment of habitation and storage rooms suggests that storage was private or restricted to some extent. It is likely that on a small site most people would know the contents of most store rooms (Gilman 1987:556). Furthermore, Lightfoot (1992: 242) suggests that the storage rooms within a suite of rooms (including two or three habitation rooms) associated with a single pit structure might have been interconnected. Clearly, pueblo store rooms were not private vaults. However, the attachment of the store rooms to one or a few habitation rooms suggests that not all residents of a site had equal access to all the stores.

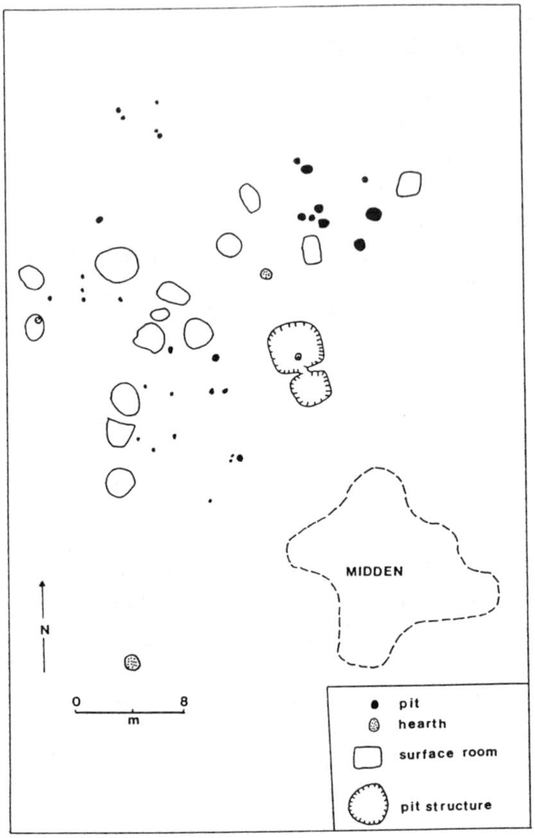

FIGURE 3 Plan of the Tres Bobos Hamlet (5MT4545), showing a single pit structure and scattered small storage structures (after Brisbin and Varien 1986:Figure 3.11).

Not surprisingly, there are some deviations from these patterns. At the SU Site, a pithouse village in the Mogollon highlands, multiple storage pits in the large pithouses suggest that storage was mostly private (Wills 1991). At some small pueblo sites (e.g., D:11:2025 on Black Mesa [Stone 1984], see Hegmon [1994]) the store rooms are not attached to the habitation rooms; thus the stores would have been relatively public. These examples make clear that the pithouse-to-pueblo transition was not the same in all times and places. However, the focus here is on the transition from more public to more private storage, whether it coincides with, precedes, or follows the development of above-ground architecture.

A great deal of research has been devoted to understanding and explaining the pithouse-to-pueblo transition and similar processes in other parts of the world (e.g., Flannery 1972; Gillespie 1976; Gilman 1987; Lipe and Breternitz 1980; McGuire and Schiffer 1983; F. Plog 1974; S. Plog 1990; Roberts 1939; Steward 1937; Whalen 1981; Wilshusen 1988). Much of this research offers multicausal explanations, often tied up

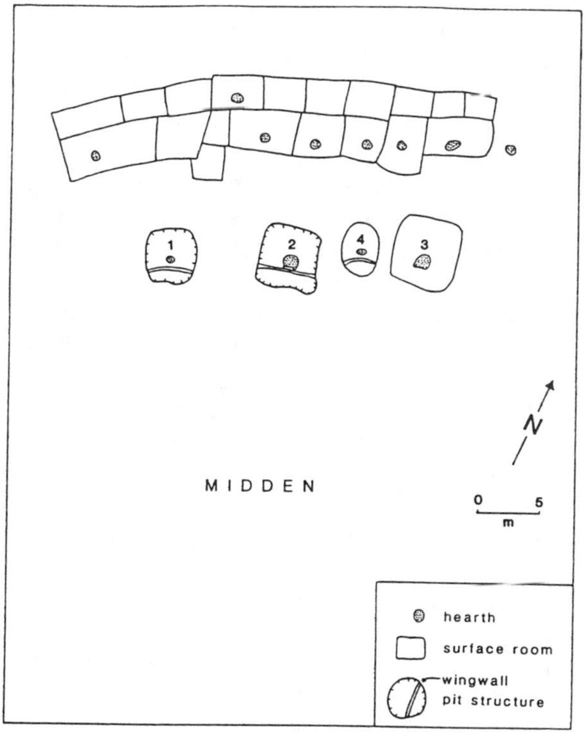

FIGURE 4 Plan of the Duckfoot Site (5MT3868), showing a double row of rooms with habitation rooms in front and storage rooms in rear (data: Lightfoot and Etzkorn 1993).

with complex social processes such as integration and differentiation. The need for a greater volume of storage space, more secure storage, and greater differentiation of stores is a part of many of these arguments. Others (e.g., Ingold 1983, 1986, 1988; Kelly 1991; Soffer 1989; Testart 1982; Young 1992) have related storage practices to subsistence and settlement strategies as well as social organization.

My goal here is to weave these various threads of argument and evidence together with the simulation results to contribute to an understanding of prehistoric economic strategies. How can we understand the change in storage strategies from somewhat public to mostly private? The simulation results, specifically the success of the restricted sharing strategy, strongly suggest that private storage would be advantageous to food producers in the variable environment of the Southwest.

This conclusion can be interpreted from various theoretical perspectives. From a processual viewpoint, the development of private storage as part of pueblo architecture can be seen as a successful adaptation for the prehistoric horticulturalists. In some cases, this adaptation may be associated with population increase and an intensification of food production, though evidence suggests that dependence on corn changed very little between the Basketmaker and Pueblo periods (Decker and Tieszen 1989; Matson and Chisholm 1991; Minnis 1989; see discussion above).

Across much of the Colorado Plateau, the new adaptation can also be seen as a response to an increase in high-frequency environmental variability beginning around A.D. 750 (Dean 1995; Dean et al. 1985).

The concept of the mode of production provides a different kind of understanding that removes focus from subsistence strategies per se. Specifically, the change from public to private storage can be interpreted as a transformation in the mode of production, that is, as a fundamental change in the social order. Why would such a transformation come about? In Marxist terms, a transformation of the mode of production is brought about as a result of tensions between the forces of production and the relations of production. In terms of the prehistoric Southwest, this means the development of private storage can be seen as a result of the incompatibility of reliance on food production and the public storage and pooling of resources.

At least two lines of evidence indicate that such tension was present and growing during late pithouse occupations. First, there are some suggestions of conflict during Basketmaker III times, including stockaded sites (e.g., Payne and Gilliland in southwestern Colorado [Rohn 1974, 1975]; possibly sites in the Gobernador District in northern New Mexico [Hall 1944]); and injuries on skeletons (e.g., at Broken Flute Cave [Morris 1980]). Second, the incompatibility of food production and pooling with communal storage may have been exacerbated by the increase in environmental variability that began around A.D. 750.

Finally, a selectionist argument (e.g., Dunnell 1980; Leonard and Abbott 1992) provides yet another perspective. Selection, whether natural or cultural, must act on variation, and Basketmaker III occupations provide much evidence of such variation. Although the bulk of the storage at pithouse villages was apparently in exterior cists and rooms, there were storage areas (including pits, antechambers, and areas behind wing walls) in individual pithouses. If restricted sharing became more advantageous, these mechanisms for private storage would have been selected, leading in part to the development of pueblo architecture.

CONCLUSIONS

The simulation model developed in this research provides a means of considering the advantages and disadvantages of various modes of production. The primary conclusion of this research is that a system of restricted sharing, in which households maintain their own stores and share only their surpluses, is far more beneficial than either a system with no sharing or a system in which all resources are pooled. Restricted sharing does not minimize variance, but by including both private storage and some sharing, it increases the chances that a household will survive a prolonged bad period. Thus restricted sharing is particularly advantageous in highly variable environments. Furthermore, restricted sharing is probably a good strategy for coping with the variability inherent in most food-production strategies.

Scholars of sedentary food-producing societies have long noted that social relations among food producers tend to involve some social differences and boundaries, and they have explained these inequalities by considering everything from land tenure systems, to intergenerational dependencies, to complex planning for delayed returns (e.g., Ingold 1983; Meillassoux 1972, 1973; Woodburn 1982). This research suggests another reason for this social differentiation. That is, social boundaries help to limit sharing and thus promote an advantageous distribution of yields among food producers.

These conclusions have implications for issues of social differences and inequality, as they have been debated recently with data from the prehistoric Southwest (e.g., Cordell et al. 1987; Reid et al. 1989). The simulation assumed a minimum of inequalities. Simulated households did better or worse than other households only because of differences in computer-generated probabilities. And even with such minimal inequalities, too much interdependence was a disadvantage. In real life, with lazy or unskilled people and poor patches of land, the disadvantages of too much interdependence would be exacerbated. Thus these results, as well as other arguments (cited above) regarding food production, lead to the rather unappealing conclusion that inequality is a beneficial strategy, at least for food producers. Furthermore, such inequality is not advantageous only to those on top of the hierarchy. Instead, inequality can be beneficial (i.e., the most successful strategy) for most members of a society. Thus inequality might emerge out of a general consensus, and need not be a result of elite control or surplus production.

ACKNOWLEDGMENTS

Many friends and colleagues contributed to this research. I am grateful to Chuck Adams, Mike Adler, Richard Ford, Carol Goland, Susan Gregg, Karin Jones, Roy Paul, Steve Plog, Scott Rushforth, Carla Sinopoli, John Speth, Heather Trigg, Mark Varien, Bruce Winterhalder, and Henry Wright. The workshop participants and organizers also provided a great deal of stimulus to this research; I am grateful to them and to the SFI and National Forest Service for hosting the conference. Joe Tainter provided useful editorial comments on this paper. Part of the simulation research was done while I was supported by a National Science Foundation Graduate Fellowship.

REFERENCES

Adams, E. Charles

1979 Cold Air Drainage and Length of Growing Season in the Hopi Mesas Area. *The Kiva* 44:285–296.

Barnard, Alan, and James Woodburn

1988 Property, Power and Ideology in Hunter-Gathering Societies: An Introduction. In *Hunters and Gatherers 2: Property, Power and Ideology*, edited by T. Ingold, D. Riches, and J. Woodburn, pp. 4–31. Berg, Oxford.

Beaglehole, Ernest

1937 *Notes on Hopi Economic Life*. Publications in Anthropology 15. Yale University, New Haven.

Bender, Barbara

1990 The Dynamics of Nonhierarchical Societies. In *The Evolution of Political Systems*, edited by Steadman Upham, pp. 247–263. Cambridge University Press, Cambridge.

Bradfield, Maitland

1971 *The Changing Pattern of Hopi Agriculture*. Occasional Paper No. 30. Royal Anthropological Institute of Great Britain and Ireland.

Braun, David P., and Stephen Plog

1982 Evolution of "Tribal" Social Networks: Theory and Prehistoric North American Evidence. *American Antiquity* 47:504–525.

Brisbin, Joel M., and Mark Varien

1986 Excavations at Tres Bobos Hamlet (Site 5MT4545): A Basketmaker III Habitation. In *Dolores Archaeological Program: Anasazi Communities at Dolores: Early Anasazi Sites in the Sagehen Flats Area*, compiled by A. E. Kane and G. T. Gross, pp. 119–210. U.S. Department of the Interior, Bureau of Reclamation, Engineering and Research Center, Denver.

Brown, William L., E. G. Anderson, and Roy Tuchawena, Jr.

1952 Observations on Three Varieties of Hopi Maize. *American Journal of Botany* 39:597–609.

Cadelina, Rowe Villaseca

1982 *Batak Interhousehold Food Sharing: A Systematic Analysis of Food Management of Marginal Agriculturalists in the Philippines*. Ph.D. dissertation, University of Hawaii.

Cashdan, Elizabeth A.

1985 Coping with Risk: Reciprocity Among the Basarwa of Northern Botswana. *Man* 20:454–474.

Collins, G. N.

1914 A Drought-Resisting Adaptation in Seedlings of Hopi Maize. *Journal of Agricultural Research* 1:293–302.

Connelly, John C.
1979 Hopi Social Organization. In *Handbook of North American Indians, Vol. 9, Southwest*, edited by A. Ortiz, pp. 539–553. Smithsonian Institution, Washington, D.C.

Cordell, Linda S., and George G. Gumerman (editors)
1989 *Dynamics of Southwest Prehistory.* Smithsonian Institution Press, Washington, D.C.

Cordell, Linda S., and Fred Plog
1979 Escaping the Confines of Normative Thought: A Reevaluation of Puebloan Prehistory. *American Antiquity* 44:405–429.

Cordell, Linda S., Steadman Upham, and S. L. Brock
1987 Obscuring Cultural Patterns in the Archaeological Record: A Discussion from Southwestern Archaeology. *American Antiquity* 52:565–577.

Dean, Jeffrey S.
1969 *Chronological Analysis of Tsegi Phaste Sites in Northeastern Arizona.* Papers of the Laboratory of Tree-Ring Research 3. Tucson, Arizona.
1992 Demography, Environment, and Subsistence Stress. This volume.

Dean, Jeffrey S., Robert C. Euler, George J. Gumerman, Fred Plog, Richard H. Hevly, and Thor N.V. Karlstrom
1985 Human Behavior, Demography, and Paleoenvironment on the Colorado Plateau. *American Antiquity* 50:537–554.

Decker, K. W., and L. Tieszen
1989 Isotopic Reconstruction of Mesa Verde Diet from Basketmaker III to Pueblo III. *The Kiva* 55:33–47.

Dunnell, Robert C.
1980 Evolutionary Theory and Archaeology. In *Advances in Archaeological Method and Theory*, vol. 3, edited by M. B. Schiffer, pp. 35–99. Academic Press, New York.

Eggan, Fred R.
1950 *Social Organization of the Western Pueblos.* University of Chicago Press, Chicago.

Evans-Pritchard, E. E.
1940 *The Nuer.* Clarendon Press, Oxford.

Flannery, Kent V.
1972 The Origins of the Village as a Settlement Type in Mesoamerica and the Near East: A Comparative Study. In *Man, Settlement and Urbanism*, edited by P.U. Ucko, R. Tringham, and G. W. Dimbleby, pp. 321–336. Duckworth, London.

Forde, C. Daryll
1931 Hopi Agriculture and Land Ownership. *Journal of the Royal Anthropological Institute of Great Britain and Ireland* 61:357–405.

Gillespie, William B.
1976 *Culture Change at the Ute Canyon Site: A Study of the Pithouse-Kiva Transition in the Mesa Verde Region.* M.A. thesis, University of Colorado.

Gilman, Patricia A.
1987 Architecture as Artifact: Pit Structures and Pueblos in the American Southwest. *American Antiquity* 52:538–565.

Green, Margerie, Keith Jacobi, Bruce Boeke, Helen L. O'Brien, Elizabeth S. Word, Richard L. Boston, Heather B. Trigg, Gilbert D. Glennie, and Melissa Gould
1985 Arizona D:11:2030. In *Excavations on Black Mesa, 1983: A Descriptive Report*, edited by A. L. Christenson and W. J. Parry, pp. 223–259. Investigations Research Paper No. 46. Center for Archaeological, Southern Illinois University, Carbondale.

Hack, John T.
1942 *The Changing Physical Environment of the Hopi Indians of Arizona.* Papers of the Peabody Museum of American Archaeology and Ethnology, Vol. 35, No. 1. Harvard University, Cambridge, MA.

Hall, Jr., E. T.
1944 *Early Stockaded Settlements in the Gobernador, New Mexico: A Marginal Anasazi Development from Basketmaker III to Pueblo I Times.* Studies in Archaeology and Ethnology II(1). Columbia, New York.

Hames, Raymond
1990 Sharing Among the Yanomamo: Part I, the Effects of Risk. In *Risk and Uncertainty in Tribal and Peasant Economies*, edited by E. Cashdan, pp. 89–105. Westview Press, Boulder.

Harris, Laurence
1983 Forces and Relations of Production. In *A Dictionary of Marxist Thought*, edited by T. Bottomore, L. Harris, V. G. Kiernan, and R. Miliband, pp. 178–180. Harvard University Press, Cambridge.

Hays, Kelley A.
1991 Social Contexts of Style and Information in a Seventh-Century Basketmaker Community. Paper presented at the 56th Annual Meeting of the Society for American Archaeology, New Orleans.

Hegmon, Michelle
1989 Risk Reduction and Variation in Agricultural Economies: A Computer Simulation of Hopi Agriculture. *Research in Economic Anthropology* 11:89–121.

1991 The Risks of Sharing and Sharing as Risk Reduction: Interhousehold Sharing in Egalitarian Societies. In *Between Bands and States*, edited by S. A. Gregg, pp. 309–329. Investigations Occasional Paper No. 9. Center for Archaeological, Southern Illinois University, Carbondale.

1994 Boundary-Making Strategies in Early Pueblo Societies: Style and Architecture in the Kayenta and Mesa Verde Regions. In *The Ancient Southwestern Community: Models and Methods for the Study of Prehistoric*

Social Organization, edited by W. H. Wills and R. D. Leonard, pp. 171–190. University of New Mexico Press, Albuquerque.

Heinen, H. Dieter, and Kenneth Ruddle

1974 Ecology, Ritual, and Economic Organization in the Distribution of Palm Starch Among the Warao of the Orinco Delta. *Journal of Anthropological Research* 30:116–138.

Hill, James N.

1970 Broken K Pueblo: Prehistoric Social Organization in the American Southwest. In *Anthropological Papers of the University of Arizona*, No. 18. University of Arizona Press, Tucson.

Himmelweit, Susan

1983 Mode of Production. In *A Dictionary of Marxist Thought*, edited by T. Bottomore, L. Harris, V. G. Kiernan, and R. Miliband, pp. 335–337. Harvard University Press, Cambridge.

Hobler, Phil

1974 The Late Survival of Pithouse Architecture in the Kayenta Anasazi Area. *Southwestern Lore* 40(2):1–44.

Ingold, Tim

1983 The Significance of Storage in Hunting Societies. *Man* 18:553–571.

1986 *The Appropriation of Nature: Essays on Human Ecology and Social Relations.* Manchester University Press, Manchester.

1988 Notes on the Foraging Mode of Production. In *Hunters and Gatherers 1: History, Evolution, and Social Change*, edited by T. Ingold, D. Riches, and J. Woodburn, pp. 269–285. Berg, Oxford.

Kane, Allen E.

1986 Prehistory of the Dolores River Valley. In *Dolores Archaeological Program: Final Synthetic Report*, compiled by D. A. Breternitz, C. K. Robinson, and G. T. Gross, pp. 353–435. U.S. Department of the Interior, Bureau of Reclamation, Engineering and Research Center, Denver.

Kaplan, Hillard, and Kim Hill

1985 Food Sharing Among Ache Foragers: Tests of Explanatory Hypotheses. *Current Anthropology* 26:223–246.

Kelly, Robert L.

1991 Sedentism, Sociopolitical Inequality, and Resource Fluctuations. In *Between Bands and States*, edited by S. A. Gregg, pp. 135–158. Investigations Occasional Paper No. 9. Center for Archaeological, Southern Illinois University, Carbondale.

Kennard, Edward A.

1979 Hopi Economy and Subsistence. In *Handbook of North American Indians, Vol. 9, Southwest*, edited by A. Ortiz, pp. 554–563. Smithsonian Institution, Washington, D.C.

Leacock, Eleanor, and Richard Lee

1982 Introduction. In *Politics and History in Band Societies*, edited by E. Leacock and R. Lee, pp. 1–20. Cambridge University Press, Cambridge.

Lee, Richard B.
1988 Reflections on Primitive Communism. In *Hunters and Gatherers 1: History, Evolution, and Social Change*, edited by T. Ingold, D. Riches, and J. Woodburn, pp. 252–268. Berg, Oxford.

Leonard, Robert, and Alysia Abbott
1992 Theoretical Aspects of Subsistence Stress and Cultural Evolution. Paper presented at SFI workshop on Resource Stress, Economic Uncertainty, and Human Response in the Prehistoric Southwest, Santa Fe, NM, 1992.

Lightfoot, Ricky R.
1992 *Archaeology of the House and Household: A Case Study of Assemblage Formation and Household Organization in the American Southwest.* Ph.D. Dissertation, Washington State University. University Microfilms, Ann Arbor.

Lightfoot, Ricky R., and Mary C. Etzkorn (editors)
1993 *The Duckfoot Site: Volume I: Descriptive Archaeology.* Occasional Paper No. 3. Crow Canyon Archaeological Center, Cortez, Colorado.

Lipe, William D., and Cory D. Breternitz
1980 Approaches to Analyzing Variability Among Dolores Area Structures, A.D. 600–950. *Contract Abstracts and CRM Archaeology* 1(2):21–28.

Matson, R. G., and Brian Chisholm
1991 Basketmaker II Subsistence: Carbon Isotopes and Other Dietary Indicators from Cedar Mesa, Utah. *American Antiquity* 56:444–459.

McGuire, Randall H.
1993 Archaeology and Marxism. In *Archaeological Method and Theory*, Vol. 5, edited by M. B. Schiffer, pp. 101–157. University of Arizona Press, Tucson.

McGuire, Randall H. and Michael B. Schiffer
1983 A Theory of Architectural Design. *Journal of Anthropological Archaeology* 2:277–303.

Meillassoux, Claude
1972 From Reproduction to Production. *Economy and Society* 1:93–105.
1973 On the Mode of Production of the Hunting Band. In *French Perspectives in African Studies*, edited by P. Alexandre, pp. 187–203. Oxford University Press, London.

Minnis, Paul E.
1989 Prehistoric Diet in the Northern Southwest: Macroplant Remains from Four Corners Feces. *American Antiquity* 54:543–563.

Morris, Elizabeth Ann
1980 Basketmaker Caves in the Prayer Rock District, Northeastern Arizona. In *Anthropological Papers of the University of Arizona*, No. 35. University of Arizona Press, Tucson.

O'Shea, John
1981 Coping with Scarcity: Exchange and Social Storage. In *Economic Archaeology*, edited by A. Sheridan and G. N. Bailey, pp. 167–183. British Archaeological Reports International Series 96.

Plog, Fred
1974 *The Study of Prehistoric Change.* Academic Press, New York.
1978 The Keresan Bridge: An Ecological and Archaeological Account. In *Social Archaeology*, edited by C. Redman et al., pp. 349–372. Academic Press, New York.

Plog, Stephen
1990 Agriculture, Sedentism, and Environment in the Evolution of Political Systems. In *The Evolution of Political Systems*, edited by Steadman Upham, pp. 177–199. Cambridge University Press, Cambridge.

Price, T. Douglas, and James A. Brown (editors)
1985 *Prehistoric Hunter-Gatherers: The Emergence of Cultural Complexity.* Academic Press, New York.

Real, L., and T. Caraco
1986 Risk and Foraging in Stochastic Environments. *Annual Review of Ecology and Systematics* 17:371–390.

Reid, J. Jefferson, Michael B. Schiffer, Stephanie M. Whittlesey, Madeleine J. Hinkes, Alan P. Sullivan III, Christian E. Downum, William A. Longacre, and H. David Tuggle
1989 Perception and Interpretation in Contemporary Southwestern Archaeology: Comments on Cordell, Upham, and Brock. *American Antiquity* 54:802–814.

Roberts, Frank H. H.
1939 The Development of a Unit-Type Dwelling. In *So Live the Works of Man*, edited by D. D. Brand and F. E. Harvey, pp. 311–323. University of New Mexico Press, Albuquerque.

Rohn, Arthur H.
1974 Payne Site Investigations. *Southwestern Lore* 40(3&4):50–52.
1975 A Stockaded Basketmaker III Village at Yellow Jacket, Colorado. *The Kiva* 40:113–119.

Roumasset, James A.
1976 *Rice and Risk: Decision Making Among Low Income Farmers.* North Holland Publishing Company, Amsterdam.

Sahlins, Marshall D.
1972 *Stone Age Economics.* Aldine, Chicago.

Sellers, William D., and Richard H. Hill
1974 *Arizona Climate 1931–1972.* University of Arizona Press, Tucson.

Smith, Eric
1988 Risk and Uncertainty in the "Original Affluent Society": Evolutionary Ecology of Resource-Sharing and Land Tenure. In *Hunters and Gatherers*

1: History, Evolution, and Social Change, edited by T. Ingold, D. Riches, and J. Woodburn, pp. 222–251. Berg, Oxford.

Soffer, Olga

1989 Storage, Sedentism and the Eurasian Paleolithic Record. *Antiquity* 63:719–732.

Spielmann, Katherine A.

1982 *Intersocietal Food Acquisition Among Egalitarian Societies: An Ecological Analysis of Plains-Pueblo Interaction in the American Southwest.* Ph.D. dissertation, University of Michigan. University Microfilms, Ann Arbor.

Spriggs, Matthew (editor)

1984 *Marxist Perspectives in Archaeology.* Cambridge University Press, Cambridge.

Stephens, D. W., and Eric L. Charnov

1982 Optimal Foraging: Some Simple Stochastic Models. *Behavioral Ecology and Sociobiology* 10:251–263.

Steward, Julian

1937 Ecological Aspects of Southwestern Society. *Anthropos* 32:87–104.

Stone, Glenn Davis

1984 Arizona D:11:2025. In *Excavations on Black Mesa, 1982: A Descriptive Report*, edited by D. L. Nichols and F. E. Smiley, pp. 193-208. Research Paper No. 39. Center for Archaeological Investigations, Southern Illinois University, Carbondale.

Testart, A.

1982 The Significance of Food Storage Among Hunter-Gatherers: Residence Patterns, Population Densities, and Social Inequalities. *Current Anthropology* 25:523–537.

Titiev, Mischa

1938 *Dates of Planting at the Hopi Indian Pueblo of Oraibi.* Museum Note 11(5). Museum of Northern Arizona, Flagstaff.

1944 Old Oraibi: a Study of the Hopi Indians of Third Mesa. In *Papers of the Peabody Museum of American Archaeology and Ethnology 22(1).* Harvard University, Cambridge, MA.

Trigger, Bruce G.

1993 Marxism in Contemporary Western Archaeology. In *Archaeological Method and Theory,* Vol. 5, edited by M. B. Schiffer, pp. 159–200. University of Arizona Press, Tucson.

United States Department of Commerce

1972 Climatological Data 1932–1972. National Oceanic and Atmospheric Administration, Environmental Data Service, Washington, D.C.

Van West, Carla R., and Timothy A. Kohler

1992 A Time to Rend, a Time to Sew: New Perspectives on Northern Anasazi Sociopolitical Development in Later Prehistory. Paper presented at the conference on The Anthropology of Human Behavior Through Geographic Information and Analysis, Santa Barbara.

Wetterstrom, Wilma E.

1976 *The Effects of Nutrition on Population Size at Pueblo Arroyo Hondo.* Ph.D. dissertation, University of Michigan. University Microfilms, Ann Arbor.

Whalen, Michael E.

1981 Cultural-Ecological Aspects of the Pithouse-to-Pueblo Transition in a Portion of the Southwest. *American Antiquity* 46:75–92.

Wharton, Jr., Clifton R.

1971 Risk, Uncertainty, and the Subsistence Farmer: Technological Innovation and Resistance to Change in the Context of Survival. In *Studies in Economic Anthropology*, edited by G. Dalton, pp. 151–178. Anthropological Studies No. 7. American Anthropological Association, Washington, D.C.

Whiteley, Peter M.

1985a Unpacking Hopi "Clans": Another Vintage Model Out of Africa. *Journal of Anthropological Research* 41:359–374.

1985b Unpacking Hopi "Clans" II: Further Questions About Hopi Descent Groups. *Journal of Anthropological Research* 42:9–79.

Wills, W. H.

1991 Organizational Strategies and the Emergence of Prehistoric Villages in the American Southwest. In *Between Bands and States*, edited by S.A. Gregg, pp. 161–180. Occasional Paper No. 9. Center for Archaeological Investigations, Southern Illinois University, Carbondale.

1992 Plant Cultivation and the Evolution of Risk-Prone Economies in the Prehistoric American Southwest. In *Transitions to Agriculture in Prehistory*, edited by A. B. Gebaur and T. D. Price, pp. 153–176. Prehistory Press, Madison, Wisconsin.

Wills, W. H., and Thomas C. Windes

1989 Evidence for Population Aggregation and Dispersal During the Basketmaker III Period in Chaco Canyon. *American Antiquity* 54:347–369.

Wilshusen, Richard H.

1988 The Pithouse-to-Pueblo Transition: An Alternative to McGuire and Schiffer's Explanation. In *Dolores Archaeological Program: Supporting Studies: Additive and Reductive Technologies*, edited by E. Blinman, C. J. Phagan, and R. H. Wilshusen, pp. 703–708. U.S. Department of the Interior, Bureau of Reclamation, Engineering and Research Center, Denver.

Winterhalder, Bruce

1986a Diet Choice, Risk, and Food Sharing in a Stochastic Environment. *Journal of Anthropological Archaeology* 5:369–392.

1986b Optimal Foraging: Simulation Studies of Diet Choice in a Stochastic Environment. *Journal of Ethnobiology* 6:205–223.

1990 Open Field, Common Pot: Harvest Variability and Risk Avoidance in Agricultural and Foraging Societies. In *Risk and Uncertainty in Tribal and Peasant Economies*, edited by E. Cashdan, pp. 67–87. Westview Press, Boulder.

Woodburn, James
1982 Egalitarian Societies. *Man* 17:431–451.
Young, Lisa C.
1992 Pits, Rooms, Baskets, Pots: Storage Among Southwestern Farmers. Paper presented at the Southwest Symposium, Tucson, Arizona.

Linda S. Cordell
University Museum, University of Colorado, Boulder, CO 80309-0218

Models and Frameworks for Archaeological Analysis of Resource Stress in the American Southwest

The workshop *Resource Stress, Economic Uncertainty, and Human Response in the Prehistoric Southwest*, focused on a general topic of long-term interest. The organizers of the workshop deliberately sought younger scholars who are currently directing field projects and who are using or developing innovative methods of analysis in their work. The organizers hoped that the assembled group would bring fresh ideas for future research directions. That goal was indeed accomplished. Although there was no conscious attempt to represent all, or even most, of the theoretical positions currently espoused by archaeologists (see Cordell 1994), there was also no intention to admit only one or a few points of view.

Two elements of the workshop, the innovative nature of the research that was shared and the variety in perspectives represented, were perceived as congenial to the interests of the Santa Fe Institute which is an institution devoted to the study of many kinds of complex adaptive systems, including human cultures (Gell-Mann 1991). The scope and direction of the Santa Fe Institute, have been described by others who are both better informed about its activities and more articulate than I (e.g., Gell-Mann 1991; Lewin 1992). Nevertheless, there are additional points on which this workshop and the institute are in accord. The workshop participants made use of the rich and detailed paleoenvironmental and archaeological records available for the Southwest, an endeavor that has been of interest to the Santa Fe

Institute since about 1990, when it co-hosted an advanced seminar on cultural evolution in the Southwest jointly with the School of American Research (Gumerman 1994). Further, some of the models developed by the workshop participants would be amenable to some of the kinds of computer modeling that is currently a great strength of SFI. These observations are clarified below in the context of reviewing the workshop's accomplishments.

The workshop organizers selected the topic *Resource Stress, Economic Uncertainty, and Human Responses in the Prehistoric Southwest.* On the first day of the workshop George Gumerman and Murray Gell-Mann expanded on the topic, expressing an interest in looking at resource depletion, particularly anthropogenic resource depletion. They suggested that the data of Southwest archaeology might allow an unusual and uniquely valuable long-term view of the relationship between resource depletion and cultural stability and change. Given the breadth of this working title, participants might have done a great variety of things. It is interesting to reflect on what was not done and what was done to see what themes emerged during the week. The themes, in turn, suggest both an organization for the current volume as well as direction that future research may take.

Given the title of the workshop and the focus on resource stress, there were some curious omissions from the discussions, at least from the archaeological point of view. For example, little time was spent trying to define stress, or resource stress.[1] Most of the papers, especially those by Kohler and Van West, Minnis, and Nelson, present working definitions of stress, but other papers do not. The oversight is somewhat perplexing in that within the past dozen years, very large research projects such as the Pajarito Plateau Archaeological Project (PARP) were devoted to examining stress, responses to stress, and especially ways in which stress might be reflected in the archaeological record. Although results of PARP research are being reported in the archaeological literature (Orcutt 1991), there was no indication at the workshop that measures of stress developed by PARP are among those that archaeologists routinely use.

The kinds of working definitions of stress that are reflected in the papers brought together here include decline in crop yields caused by a decrease in precipitation and specific nutritional deficiencies that may have been brought about by climatic change or game depletion as a result of overhunting. Rather than devote a great deal of time to defining stress, workshop participants did focus attention on models of economic risk and uncertainty. Kohler and Van West, and Hegmon define risk to refer to variance in (agricultural) production, rather than the probability of falling below a certain defined level. On the other hand, Nelson (this volume) defines risk, following Winterhalter (1990), as potentially harmful shortfalls resulting from unpredictable or stochastic environmental factors. Nelson's definition is probably more common among the participants, but there was no discussion of how different definitions might influence the types of models of human response to risk that we develop.

[1] See pp. 15–16.

Another topic that was omitted was the effect of stress on human physiology and biological functioning. There was no discussion of instances of apparent nutritional deprivation or elevated mortality or morbidity reflected in skeletal series. In part, this lack reflects the subfield composition of the participants, who are all archaeologists rather than biological anthropologists or osteologists. This subfield specialization is also reflected in the lack of discussion concerning long-term effects of inadequate calories or nutrients on human reproduction, although this is mentioned in the paper by Spielmann and Angstadt-Leto. Another probable reason for the neglect of this topic is that there was a general belief that most instances of resource stress were accommodated through behavioral modifications before they could cause the kinds of biological changes that would be observable on human skeletons.

Most southwestern archaeologists seem to accept the notion that because of the region's aridity and variability in the timing of rainfall, Precolumbian populations were always undergoing a moderate amount of stress. Many southwesternists assume that some level of subsistence stress was endemic and was tolerated. The view is that indigenous populations of the region may not have been robustly healthy, but they were not dying in unusual numbers for lack of nourishment.

The somewhat benign view that subsistence stress generally was not lethal to large numbers of people, except for children, I suspect, turns archaeological attention to the corollary assumption that southwestern populations were tremendously flexible and inventive and that they had great funds of knowledge, or what Gunn (1994) refers to as breadth of captured experience, concerning ways to counteract various subsistence shortfalls. This view puts southwestern archaeologists in the position of looking for those kinds of behaviors that successfully buffered or counteracted instances of diminished resource abundance. Concomitantly, there is a failure to examine the very real possibility that much of what people may have done could have made things worse in the short-term, or have made failure ultimately certain.

There is astonishment among many nonanthropologists that southwesternists appear not to view the thirteenth-century abandonment of large portions of the Southwest as one of history's most extraordinary examples of social collapse and failure. For the anthropologists, the emphasis is on the undeniable historical continuity between the Anasazi and Mogollon, and the modern Pueblos. Emphasizing continuity entails minimizing events, such as the abandonment, rather than portraying them as a failure of the human population to withstand stress. Understandably, there was no discussion of the kinds of cultural strategies that may have been implemented by the twelfth- or thirteenth-century populations that could have so altered their environment or context that abandonment became essential. It is worth noting, in this regard, that there is as yet no consensus among southwestern archaeologists about whether or not abandonment of the Four Corners region was accompanied by increased mortality. At another recent conference of southwestern archaeologists, there was no resolution to the question of whether the size of the population in the Southwest in 1200, prior to the abandonment, was greater, the same, or less than at 1500, just before Pueblo contact with Europeans (see Adler 1995).

In looking at the many issues that were discussed productively during the workshop, a few themes stand out for crosscutting different approaches. For example, in many of the papers, attention is focused on the individual or the household and on local situations. The kinds of stress-reducing or risk-reducing behaviors discussed involve strategies that are generally implemented at the household level. One implication of this perspective is that social hierarchies and social complexity (in the anthropological sense of ranked or stratified societies) is irrelevant to models of decision making in the Precolumbian Southwest. In other words, regional social dynamics and hierarchical systems of social control are not considered important to the kinds of strategies implemented in response to stress.

This view is fascinating to me for at least two reasons. It reveals that despite more than a decade of discussing the Chaco Phenomenon and the possibility of the existence of regionally based systems in the fourteenth century, most archaeologists are probably not convinced that these had any meaningful impact on local behaviors. Second, this perspective excludes considering the kinds of models developed in regions where ancient complex societies certainly existed (such as the Valley of Mexico, the Mayan area, or the Near East) that implicate the divergent interests of rulers and ruled among reasons for societal collapse and failure. Even Hegmon (this volume), who is concerned with the development of social inequality and who considers dependence on agriculture as being at its root, presents a model in which the household is the basic economic unit.

Clearly related to the above are the issues of scale that were addressed in several papers. Thus paleoclimatic patterns, such as were explored by Jeffrey Dean, are detectable only at the supraregional level. These, of course, had highly specific, local consequences that provided the context for the behaviors of the groups of concern. Similarly, Spielmann and Angstadt-Leto find that trade in meat among Eastern Anasazi and Plains groups did not develop until the fifteenth century, when the regional population had shifted to the Rio Grande and Plains margins area. In this case, change in the distribution of population at the regional scale becomes a precondition providing the social context for explaining behavior options studied at the scale of the locality. Kohler and Van West also suggest that regional population density will modify the utility function they propose for the value of resource pooling. Dean's discussion of local responses to the environment depends upon both regional population densities and regional climate patterns.

Another workshop theme was reciprocal obligations. In this case, the explicit anthropological assumption is that the societies under study were egalitarian and that economic exchanges were dominated by reciprocity in the technical anthropological sense (Sahlins 1968).[2] For example, Spielmann and Angstadt-Leto look

[2] For the nonanthropologist, egalitarian societies are those in which social roles are determined by age, sex, and personal qualities only. Among egalitarian societies, the number of status positions is equal to the number of individuals capable of filling them. There are no societies in which

at the potential of reciprocal exchanges between hunters and horticulturists to even out nutritional and caloric imbalances suffered by either group. Rautman explores rainfall records on the eastern margins of the Pueblo world for patterns of complementarity and therefore mutually beneficial reciprocity among horticultural. Hegmon addresses exchange from a slightly different perspective. She assumes that reciprocal exchanges were required but that if restricted sharing were a more successful strategy, then methods of excluding households from participating in exchange would have been valuable.

Along with the general focus on the household, workshop participants did not avoid discussion of conscious choices made by individuals in the populations under study. Instead, there was concern for trying to determine what perceptions actually trigger behavioral responses. Such questions consider whether people activate responses when the situation is perceived as being severe or whether they modify their behavior because a situation is viewed as being highly unusual. A related question is whether perceptions of either severity or novelty call forth different behavioral strategies. These are themes addressed in both Minnis's and Nelson's papers. Hegmon makes an important contribution to these concerns in her observation that scarcity should be felt first, before it actually causes hunger, in the social or ritual realm when there is a lack of produce needed for ritual obligations.

Collectively, the workshop participants did not attempt to sort out those environmental changes that were anthropogenic from those that were not. Yet, participants made the point that not all anthropogenic modifications reduce environmental productivity. For example, Sullivan develops the argument that some Anasazi groups used fire to increase both the abundance and diversity of edible plant foods. Minnis and Nelson note that activities that disturb the land surface, such as clearing land for fields or for building, can also increase the quantity or diversity of edible species of plants and animals.

Population aggregation emerged as another workshop theme. As has long been noted for those localities that were abandoned, aggregation, if it occurs at all, takes place immediately before abandonment. In some research such as in the PARP study area, aggregation is considered a response to stress (Hill and Trierweiler 1986; Orcutt 1991). Aggregation may facilitate some group endeavors such as cooperative hunting, as Spielmann and Angstadt-Leto and others at the workshop noted. Aggregation may also facilitate resource pooling, an observation that is the starting point for the model developed by Kohler and Van West.

Finally, although not precisely a theme, most of the papers and discussion during the workshop were informed by the tremendously rich ethnographic record

everyone is equal. Among egalitarian societies, exchange of goods is generally reciprocal. General reciprocity describes exchange that we are familiar with within the household. That is, material is given with no expectation of an equal return. In balanced reciprocity, goods of equivalent value are exchanged either immediately or at established intervals. Sahlins also defined negative reciprocity as cheating or theft. Among socially stratified groups, redistribution and market exchanges are added to the inventory of exchange types.

of the Southwest. The approaches explored by Minnis, Nelson, and Hegmon, for example, would seem to be virtually impossible in the many parts of the world where similarly detailed ethnographic literature is lacking. That the modeling efforts developed by Kohler and Van West, Nelson, and Hegmon are as convincing and robust as they are can also be related, in part, to the quality of the ethnographic record.

In looking at all the themes mentioned, the advantages of the Southwest as a laboratory are readily apparent. The remarkably long and precise record of paleoclimate and environment is matched by incomparable detail of ethnographic recording. The kinds of questions southwestern archaeologists ask routinely and the high level of resolution they expect of their data would make grown practitioners working in other parts of the world weep with jealousy. As I think these papers show, the kinds of models developed for the Southwest are detailed and sophisticated. They encourage evaluation of general ideas from anthropology and other sciences.

The papers delivered and discussed during the workshop may be divided into two broad categories that are here termed "Dynamic Models" and "Frameworks." Generally, dynamic models include mechanical or computer simulations, or mathematical models. Dynamic models are presented in the papers by Kohler and Van West, Hegmon, and Rautman. These papers present models in that they are concerned with abstracting a few qualities or dimensions of the information available and proposing ways in which they are related. The models are dynamic in that they concern processes that occur over time and portray changes over time as important elements of their design. The models are general in that they depend on basic ideas about the way certain kinds of cultural behaviors are thought to work and apply these ideas to specific situations, which, in turn, allows partial evaluation of the ideas in question.

The model that Kohler and Van West develop derives from the microeconomic theory of utility functions. This model is also considered valuable in evolutionary ecology. Kohler and Van West regard risk as year-to-year variability in crop production that is a result of climatic fluctuation. They note that in several verbal, nonmathematical descriptions in Southwest archaeology, a common response to this kind of risk is thought to be the pooling or sharing of resources among households. Kohler and Van West argue that whether or not the Anasazi selected this strategy would depend upon whether or not the gain from pooling was perceived as greater or less than that from not sharing. The perception, they suggest, is in turn based on a utility function of sigmoidal form.

That such a model might be tested using archaeological information is remarkable and, again, a tribute to the quality of the data with which southwesternists work. Kohler and Van West can and do model crop production under retrodicted climatic conditions over a series of years for a particular area. The surrogate measure they use for food sharing is population aggregation, which they measure archaeologically by looking at the sizes of sites. Sharing resources among households, of course, is behavior that has very little visibility archaeologically. In essence, Kohler and Van West assume that resource sharing is facilitated when the bulk of the population is living in large settlements. Therefore aggregation becomes an archaeological replacement measure for sharing. Being able to date periods of aggregation with enough precision to use the crop production data derived from the annual tree-ring records of precipitation is again most unusual for archaeology.

Kohler and Van West note that the assumptions they made must be closely scrutinized, and there are questions that can be raised regarding the model. For example, does the presence of aggregated sites really indicate resource pooling or do households within such settlements remain economically independent? Kohler and Van West also incorporate information about regional population packing, suggesting that extra-local population density has the effect of changing the shape of the utility curve. Although it is clear that the utility function of concern is not linear, another question that can be raised is whether there are corroborative data that indicate that the curve is sigmoidal. What would be the predictions given utility functions of different shapes? Despite these questions, it is exciting from the perspective of more general theory that a model derived from microeconomics is relevant to evolutionary ecology, and as useful as it is in predicting such diverse activities as the feeding strategies of juncos and the choices in resource sharing made by ancient Puebloan farmers.

Hegmon also examines agricultural risk in terms of climatic variation, and like Kohler and Van West her unit of analysis is the household. She also uses computer simulation as a modeling tool. Whereas Kohler and Van West consider the behavioral options to be pooling vs. nonpooling of resources among households, Hegmon's model introduces the further choice of restricted sharing, defined as sharing only surplus production above the needs of the household. Hegmon's model also includes storage, which of course is considered the particular advantage of agricultural produce in the Southwest.

Hegmon's results, like those obtained in the studies of optimal foraging strategies, microeconomics, and Kohler and Van West's study, indicate that achieving a lower variance is not always advantageous. Rather, patterns in the climatic variation were more important to the success of a particular strategy. Hegmon's study showed that the strategy that allowed surviving over a *series* of bad years was the most successful, and this proved to be restricted sharing.

Like Kohler and Van West, Hegmon links her model to pueblo settlement form, but her interest is in the change from pithouse to above-ground pueblo architecture, rather than from dispersed to aggregated pueblo settlements. Hegmon cites architectural studies of spatial relationships within pithouse villages and pueblos that support interpretations of limited access to storage in pueblos. Her findings, as she notes, have implications for understanding other aspects of Pueblo culture, such as the implementation of social boundary mechanisms.

Rautman's study also develops a dynamic model. Her interest is in exchange relationships as strategies for reducing the consequences of crop failure. Rautman explores the importance of economic exchanges from one locality to another, and the formation of social networks that facilitate these interactions. In terms of the diverse strategies that can reduce subsistence stress, maintaining extra-local social networks is, energetically, a relatively expensive solution (see Minnis, this volume). The kinds of networks Rautman describes are derived primarily from the literature of hunter-gatherers or groups practicing a minimal amount of horticulture. Although for purposes of her model that literature is appropriate, the kinds of interactions she describes are, if anything, even more important among people with greater investments in farming. Population densities among hunters and gatherers are much lower than among farmers. Extra-local social ties among hunters and gatherers often facilitate movement of individuals or households rather than formalized exchanges of economic goods. The horticultural Enga, whom Minnis discusses, are an excellent case of the use of social ties to facilitate both migrations and exchange.

Rautman's model is concerned with predicting the shape of the social network of interaction. She expects that social ties will unite groups occupying areas that are unlikely to be experiencing stress at the same time. Using the Kite site, which she excavated, as the point of reference, and modern climate records, she proposes that residents of the site would form exchange relationships with areas that are close by and also regularly experience different climatic conditions. Stated in simplified form, at times when the residents of the Kite site were in the midst of a drought, ties would be expected to be maintained with the closest group that at those times normally experienced better than average precipitation. Rautman evaluates the success of her model by using ceramic style as a measure of interaction. Those localities that do complement one another in terms of climate should exhibit more similarities in ceramic assemblages than they do with areas that experience the same climate at the same time. She finds that for her study area in central eastern New Mexico, this is the case.

Rautman suggests that her model is most useful in helping archaeologists explain the inclusion of specific sites or localities in networks defined on the basis of stylistic similarities. In fact, for many locations on the Colorado Plateaus, the paleoclimatic record based on tree rings would allow predicting when and where such networks might develop. It would then be possible to examine changes in such networks over time.

There are interesting features that unite all of the studies that present dynamic models. In all of them, the household is the basic unit of analysis. Choices are made at the level of the household based on internal assessments and information from the broader natural and social environments. In terms of the kinds of computer models developed at the Santa Fe Institute, the focus on households suggests that agant-based computer simulations of these systems might be appropriate. As indicted above, all of the models make excellent use of the climatic, and where available, paleoclimatic records, but in none of them are decisions made mechanically or exclusively by reference to climate or the environment. Quite properly, all of the models are derived from general anthropological, economic, or ecological-theoretical writing. The intent in each case is to evaluate propositions derived from these sources. Again, in my view, this is an appropriate research strategy.

Another set of workshop papers develops frameworks. These are the papers by Minnis, Spielmann and Angstadt-Leto, Nelson, Sullivan, and Dean. Frameworks are systematically organized conceptual tools that direct attention to new ways of arranging information. In the physical and biological sciences, frameworks are well established, although they may be modified as a consequence of new developments in theory. The Linnean system of taxonomy in biology, and the periodic table of elements in chemistry, are classic examples of frameworks. The use of conceptual tools such as clades, or the organization of elements with reference to their mass, provided new frameworks that have become useful through the introduction of new theory and novel forms of analysis. Archaeology is young enough as a science so that frameworks are not well established, and their development can be an important contribution to further intellectual growth of the field.

Frameworks should assist us in making new or greatly refined observations and more accurate models of the world we wish to understand. It is noteworthy that looking at resource stress and economic uncertainty is relatively new for archaeology. There is no body of established conceptual tools directed toward describing such stress or for organizing discussions of responses to it. Given this situation, the papers that elaborate frameworks stand as valuable contributions to archaeology in general.

Minnis discusses the kinds of responses individuals and groups make to food scarcity. The framework he uses and elaborates on is one that Halstead and O'Shea (1989) proposed, which groups responses into four inclusive categories: mobility, diversification, physical storage, and exchange. Minnis elaborates subcategories of these and adds two new categories of response: resource conservation and economic specialization in items that can be exchanged for food.

As Minnis notes, the usefulness of developing a framework for response categories derives from the fact that responses are not used randomly by societies, and a number of authors, including himself, have developed typologies that rank responses in terms of social and/or energetic costs. Minnis also finds it useful to distinguish between what he terms *catastrophic* and *impinging* shortages. These are identified by differences in their severity, duration of onset, and novelty. Both types may lead to culture change but the change may be incremental in the case of impinging shortages, and disruptive if the shortage is catastrophic.

The points Minnis makes in his paper integrate well with models and frameworks proposed by other contributors to this volume. For example, resource storage and pooling, critical to Kohler and Van West's model, may be seen in the context of other possible solutions to resource shortages. Similarly, the distinction between high- and low-frequency environmental processes, as described and elaborated by Dean (and see also Dean 1988), can be articulated with Minnis's distinction between impinging and catastrophic shortages. Combining Dean's and Minnis's classifications could allow retrodicting behavioral responses by Colorado Plateau populations that, in turn, might suggest more detailed and refined observations than we now make. As one example, we might retrodict times when polyculture would have been an expected response to shortages, rather than productive specialization and exchange of food items.

Sullivan's contribution is unique in this volume, because it alone focuses primarily on the anthropogenic environment. Further, Sullivan discusses the likelihood that indigenous Southwesterners used fire to enhance the productivity of edible woodland understory plant foods. If verified, this observation introduces an important behavioral option that has been ignored by southwestern archaeologists. As Sullivan tells us, fire was used by Great Basin, Plateau, and California hunter-gatherers and is now understood to have been a major tool of resource management in aboriginal Australia, Europe, and Southwest Asia. We certainly need to know whether or not, and probably within what environmental limits, fire might have been used in the Southwest. It would also be useful to know how using fire in this way would be incorporated in the framework provided by Minnis.

Sullivan points out, accurately I believe, that archaeologists have been remiss in fully evaluating the contexts in which food remains (and tools associated with food) are recovered. He notes that these contexts are primarily related to food consumption rather than production. Hence, we underestimate the importance of the contexts of food production in general. Although Sullivan's paper does not develop a framework for analytical inquiry, his suggestion that we are potentially missing a component of past behavior directs us toward developing such a framework. A complete framework to examine economic production would be very useful indeed.

Spielmann and Angstadt-Leto direct our attention to the nutritional value of components of prehistoric diets, in addition to the more usual consideration of caloric or protein values. They consider meat shortages that might have occurred among sedentary, aggregated villagers, and strategies that could have alleviated the associated nutritional deficiencies. The strategies they describe are trade in meat, turkey husbandry, and harvest of plants containing nutrients found in meat. These strategies are reflected differentially in the archaeological record of the Southwest.

The framework focus on nutrition is important. It directs us to consider the results of resource stress measured by factors other than calories and protein values. Second, it requires the further development and more consistent application of bioarchaeological analyses currently available. Although not mentioned by Spielmann and Angstadt-Leto, it appears that archaeological ubiquity and abundance measures for plants are not as well developed as they are for faunal remains, and it may be difficult to interpret minor amounts of some plants or their absence from archaeological contexts. For example, beans are underrepresented archaeologically because processes that make them palatable (such as soaking) also encourage their decay, and the plant parts that might be preserved, because they are relatively hard, are also consumed (Minnis 1985). If pollen rather than macrobotanical remains are the focus of examination, the problems can be further complicated by differential pollen production or resistance to decay. Extraordinary methods of extraction and counting may be necessary (cf. Dean 1992). Nevertheless, Spielmann and Angstadt-Leto clearly document the value of pursuing research guided by human nutritional requirements, and the measurement problems can be overcome. Having mentioned them might inspire some work in that direction.

The first part of Jeffrey Dean's paper is a thoughtful essay on the utility of certain concepts used by faculty associated with the Santa Fe Institute to study the evolution of complex adaptive systems, his own use of the concept of adaptation, and why examining adaptation does not preclude the role of selection. Dean also explores the meaning of the terms adaptation and evolution. This part of Dean's discussion reminds us that there are important differences in the way genetic and "cultural" information is transmitted and maintained in human populations (cf. Boyd and Richerson 1985; Cavalli-Sforza and Feldman 1981). Dean then turns to identifying processes in the natural and cultural environments that provide the context for behaviors that are subject to selection. These, in turn, provide a framework for structuring research.

Dean explores the complicated interrelationships among patterns in natural and cultural environmental variability, human demography, and human behavior that can be teased out of the archaeological and paleoenvironmental data from the Southwest. Of particular interest is the observation that it is the structure of relationships among *patterns* of variables that is crucial, rather than those among the variables themselves. Dean's case study is the relationship among patterns in the environmental, demographic, and behavioral variability over the past 2,000 years in the Southwest, enhanced by the long and detailed tree-ring record. Dean reports on his most recent research that shows a long-term patterned relationship between precipitation records, derived from tree rings, which contrast the northwest and southeast portions of the region in general. This stable, long-term pattern breaks down for a 200-year interval between A.D. 1250 and 1450, a period southwestern archaeologists immediately recognize as corresponding with the abandonment of the Four Corners, the development of aggregated settlements, and the major dislocations of populations across the landscape. A. E. Douglass, founder of the science of dendrochronology, published an article in 1929 claiming in its title, "The Secret of the Southwest Solved by Talkative Tree-Rings"; should we call Dean's paper "Part II"? Perhaps one day we will. Meanwhile, the opportunity to evaluate Dean's discovery is available, in part, through models of world climate (see Gunn 1994). This evaluation promises to create a period of very exciting research.

There are at least two additional topics raised by Dean that would reward additional research efforts. As it stands, Dean's (1988) model may underestimate the potential range of human behavioral responses that may have mitigated the effects of climatic perturbations. In addition to increasing storage, such behaviors might have included all of the activities Minnis discusses, diverse technologies of field selection, planting strategies, construction of water and soil control features, and perhaps, following Sullivan, using fire to clear understory plants. Second, while the distinction Dean makes between short-term environmental processes and long-term environmental processes seems to be very useful, additional research might be directed toward refining the cut-off of the suggested 25 years. The use of 25 years as a surrogate for one human generation is reasonable but may obscure great variability that might be patterned in interesting ways. If behavioral responses to stress develop and are maintained as part of cultural memory as a product of the temporal depth of capturing experiences of environmental changes, as Gunn (1994) suggests, then the length of time over which change is understood should vary among peoples inhabiting different environments. This general question is one in which there is on-going research by anthropologists and archaeologists working in different parts of the world.

The chapter by Nelson clearly defines risk and the social and economic decisions that could be implemented with the intention of reducing risk. The framework she develops is explicitly concerned with technological strategies that would result if specialization and diversification, which are the two most important economic strategies of risk reduction, were implemented. Considering that technology, including both knowledge and tools, is frequently at the center of causal models in evolutionary anthropology (e.g., White 1949), and that of the three major components of cultural systems (ideological, sociological, and technological), technology is the most salient archaeologically, it is remarkable that so little systematic discussion has been focused on the analysis of technological systems. Nelson's paper is an outstanding, thoughtful movement toward overcoming this lapse. The framework she develops is informed by an immense amount of literature and relevant personal observations. It is guided by the concept of technological organization, which she defines as the relationships among strategies for manufacturing, manipulating, and abandoning material items. It is then systematically applied to the general strategies related to resource specialization and diversification as they would be reflected in the subsistence-related tasks practiced among the prehistoric peoples of the Southwest.

The framework goes far beyond a useful classification. It allows us to appreciate the kinds of tradeoffs and compromises that must have been made when behaviors were modified or new ones were adopted. It encourages our understanding of the complex ramifications of decisions that were made every time a tool or facility was made and used. Nelson uses the framework to imagine changes that would have been made in hunting technology by Southwestern peoples experiencing subsistence risk, and developing either more specialized or diversified hunting strategies. Finally, Nelson provides an evaluation of the value of the framework as a conceptual tool by examining the design of hunting weaponry in five prehistoric southwestern contexts, with results that are both informative and novel.

It is clear that a great deal of thought and effort went into all of the models and frameworks presented at the workshop. Archaeologists working in the Southwest will find stimulating thoughts throughout the volume, as will those who are concerned, in a more general way, with long-term changes in resources and human responses to variability in resources. For these individuals, the quality of the "Southwest archaeological laboratory," as demonstrated by the innovative approaches used by the contributors, may suggest new avenues for their own research.

It is also possible to suggest directions for future research that could reflect Santa Fe Institute interests and approaches. As noted above, the dynamic models presented here are concerned with decisions made at the individual or household level. This scale provides a baseline and suggests that developing agent-based computer simulations could be an appropriate phase of investigation. For example, such computer models might examine the results of aggregates of communities, on model landscapes of various sizes, following the strategies outlined in Kohler and Van West's and Hegmon's simulations. Another possibility would be to develop simulations that consider household-level decisions based on utility functions for several resources, such as cultivated crops, game, and wild plant foods.

The frameworks developed by Minnis, Spielmann and Angstadt-Leto, Sullivan, and Nelson ultimately concern decision making that involves compromises and ranking of desirable outcomes. All of them are amenable to modeling through a genetic algorithm (Holland 1975; Forrest 1993), and actually doing so should be a very rewarding endeavor. A genetic algorithm would track how a system might learn and perfect a series of responses to impinging or catastrophic shortages or both. Or, a genetic algorithm could learn to make the least-energetically costly solution to a very realistic scenario of resource variation. Other options are easy to imagine, although all of them would be difficult to program. The effort, I suspect, would be worth the investment. Dean's framework was actually suggested in an earlier publication (Dean 1988), and it has been used and referenced by a number of other researchers, including those whose papers are included here (see Cordell and Gumerman 1989). The recent work Dean has done on patterns of regional climate change are properly incorporated in on-going research and further analysis.

Finally, the workshop papers as a group reflect a sense of the excitement that develops among scholars when they share a sense of problem. This inspiration, coupled with thorough familiarity with the remarkable data of Southwest archaeology, makes for an excellent and useful conference volume.

REFERENCES

Adler, Michael (editor)

1995 *Pueblos in Transition: The Anasazi World, A.D. 1100–1400.* University of Arizona Press, Tucson (in press).

Boyd, Robert, and Peter J. Richerson

1985 *Culture and the Evolutionary Process.* University of Chicago Press, Chicago.

Cavalli-Sforza, L. L., and Marcus W. Feldman

1981 *Cultural Transmission and Evolution: A Quantitative Approach.* Princeton University Press, Princeton.

Cordell, Linda S., and George J. Gumerman (editors)

1989 *Dynamics of Southwest Prehistory.* Smithsonian Institution Press, Washington, D.C.

Cordell, Linda S.
1994 The Nature of Explanation in Archaeology: A Position Statement. In *Understanding Complexity in the Prehistoric Southwest*, edited by George J. Gumerman and Murray Gell-Mann, pp. 149–162. SFI Studies in the Sciences of Complexity, Proc. Vol. XIV. Addison Wesley, Reading.

Dean, Glenna
1992 In Search of the Rare: Pollen Evidence of Prehistoric Agriculture. Paper delivered at the New Mexico Archaeological Council Agricultural Symposium, October 2–4, 1992, Santa Fe.

Dean, Jeffrey S.
1988 A Model of Anasazi Behavioral Adaptation. In *The Anasazi in a Changing Environment*, edited by George J. Gumerman, pp. 25–44. Cambridge University Press, Cambridge.

Douglass, Andrew E.
1929 The Secret of the Southwest Solved by Talkative Tree-Rings. *The National Geographic Magazine* 56(2):736–770.

Forrest, Stephanie
1993 Genetic Algorithms: Principles of Natural Selection Applied to Computation. *Science* 261:872–878.

Gell-Mann, Murray
1991 *The Santa Fe Institute.* SFI Working Paper 91-03-107, Santa Fe Institute, Santa Fe.

Gumerman, George J. (editor)
1994 *Themes in Southwest Prehistory.* School of American Research Press, Santa Fe.

Gunn, Joel D.
1994 Global Climate and Regional Biocultural Diversity. In *Historical Ecology: Cultural Knowledge and Changing Landscapes*, edited by Carole L. Crumley, pp. 67–99. School of American Research Press, Santa Fe.

Halstead, Paul, and John O'Shea
1989 Introduction. In *Bad Year Economics: Cultural Responses to Risk and Uncertainty*, edited by Paul Halstead and John O'Shea, pp. 1–5. Cambridge University Press, Cambridge.

Hill, James N., and W. Nicholas Trierweiler
1986 *Prehistoric Responses to Food Stress on the Pajarito Plateau, New Mexico.* Technical Report and Results of the Pajarito Archaeological Research Project, 1977–1985. Ms. report submitted to the National Science Foundation, Washington, D.C.

Holland, John H.
1975 *Adaptation in Natural and Artificial Systems.* University of Michigan Press, Ann Arbor.

Lewin, Roger
1992 *Complexity: Life at the Edge of Chaos.* Macmillan, New York.

Minnis, Paul E.
1985 Domesticating People and Plants in the Greater Southwest. In *Prehistoric Food Production in North America*, edited by Richard I. Ford, pp. 309–339. Anthropological Papers No. 75. Museum of Anthropology, University of Michigan, Ann Arbor.
Orcutt, Janet D.
1991 Environmental Variability and Settlement Changes on the Pajarito Plateau, New Mexico. *American Antiquity* 56:315–332.
Sahlins, Marshall
1968 *Tribesmen*. Prentice-Hall, Englewood Cliffs, N.J.
White, Lesie A.
1949 *The Science of Culture*. Grove Press, New York.
Winterhalter, Bruce
1990 Open Field, Common Pot: Harvest Variability and Risk Avoidance in Agricultural and Foraging Societies. In *Risk and Uncertainty in Tribal and Peasant Economies*, edited by Elizabeth Cashdan, pp. 67–87. Westview Press, Boulder.

Index

A

B

C

D

E

I

N

O

P

Y

Z